A Graphic Guide to Industrialized Building Elements

Raymond Sluzas
Anne Ryan

A GRAPHIC GUIDE TO INDUSTRIALIZED BUILDING ELEMENTS

Raymond Sluzas

Anne Ryan

CBI

CBI Publishing Company, Inc.
51 Sleeper Street
Boston, Massachusetts 02210

Library of Congress Cataloging in Publication Data

Sluzas, Raymond, 1949 —
A graphic guide to industrialized building elements.
Bibliography: p.
Includes index.
1. Industrialized building. I. Ryan, Anne, joint author. II. Title.
TH1000.S58 693.9'7 77-13121
ISBN 0-8436-0163-9 case/0-8436-0164-7 paper

Printed in the United States of America

Cover design by David Ford

CBI Publishing Company, Inc.
51 Sleeper Street
Boston, Massachusetts 02210

To Our Parents

TABLE OF CONTENTS

PREFACE

This book gives the reader some basic data on certain elements of the construction process which may be of assistance to design. A major difficulty in the building profession is keeping up to date; there is a wealth of information available that is nearly impossible to keep abreast of. The usefulness of this book lies in the fact that it acts as a filter, describing generically types of elements, and eliminating specific, proprietary information. Once the preliminary decision is made to use one or another of the building elements described herein, then more detailed information may be sought out. But at the preliminary stages, this detailed information would be confusing, a hinderance rather than a help.

The material in this book deals with a specific facet of construction — those elements of the sequence which to one extent or another are preassembled before coming to the job site. Many of these things are taken for granted now, but not too long ago it was not so.

This book is not concerned with what the impact or the implications of these changes are (that has been done much more extensively by other authors). This book presents a clear exposition of the reality, and leaves the reader to draw any conclusions or to make any applications. What is hoped for is an increased awareness, to allow understanding, manipulation and application to take place.

The authors wish to express gratitude to a number of individuals who have contributed and assisted to make this work possible. First and foremost, the assistance and encouragement of Forrest Wilson was invaluable from conception to realization. In addition, the help of Louise Addonizio, Brian Davis, Dan Karchem, Richard Lorenti, Dan Shewbridge, and Richard Weil was greatly appreciated. Finally, the many manufacturers who supplied information and assistance are acknowledged.

INTRODUCTION

'They don't make 'em the way they used to.' Modern buildings more and more are composed of pieces which are put together before they reach the building site. At various levels of complexity these preassembled parts are a necessary part of the construction process. They are a fact of life; whatever the reasons, the designer and the builder must take them into consideration and incorporate them to one degree or another as parts of the final building. The construction process has become industrialized; in fact, the construction industry is a critical part of the national economy.

Industrialization, however, has come somewhat slower to the building art. The reasons for this are complex and beyond the scope of this work. The specific nature, characteristics, and product of the building profession would have to be investigated and analyzed. What is important is that there is a certain amount of industrialization that has taken place and its impact on the design and construction process needs investigation.

The incorporation of prefabricated building elements has slowly infiltrated the building process. In many ways it is taken for granted; it has become the norm, the standard practice. It is useful to take a look at these changes — to become aware of the implications inherent in these changes.

An understanding and an appreciation of what is available are critical to a proper response to the needs of building. Knowing the tools of the trade is mandatory to the execution of the work. The problem is to put the tools in the hands of the user. With the complexity of the industry (and of society in general), the potential participant is faced with an almost overwhelming amount of information from which to choose. The natural reaction is to stay with the known, tried-and-true material. However, this may overlook a better, more efficient solution to the problem. A shorthand summary of the available resources is desirable — a 'pattern book' from which to draw inspiration so as to generate through modification and variation a solution to the specific problem.

Some discussion of terminology is in order. Certain terms have come into general use; most are vaguely understood or have acquired connotations and shades of meaning irrelevant to the present discussion. A clarification of what these terms are interpreted to mean in the context of this book will be given.

Industrialization has already been mentioned. Industrialization is taken to mean that aspect of the manufacturing process which, by the use of technological, mass production techniques, there is an increase in productivity, efficiency, or quality of the product. This can involve a product new and different in form from the pre-industrialized model, or the creation of one basically identical. Industrialization can also involve the process of production as well as the product itself. Taylor's 'scientific management' theories are in this category and lie at the foundation of the entire Industrial Revolution.

Prefabrication is very simply the assembly of certain parts of the final building prior to its delivery to the job site. Most often this pre-assembly is in a factory of one sort or another. The scope of prefabrication can range from the large to the small. Entire buildings can be manufactured off-site and then shipped to the desired location (e.g., the mobile home). Conversely, a part of a building as seemingly simple as a window is composed of a variety of sub-pieces which are now, for the most part, assembled at a production facility. In the standard parlance, prefabrication can be subdivided into four categories based upon the scale of the unit in relationship

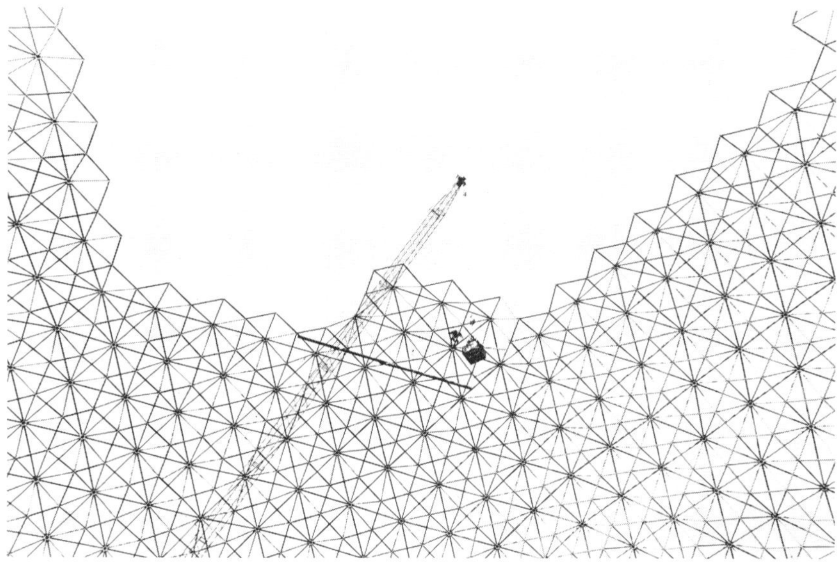

Geodesic dome at the Montreal Expo under construction

Expo Corporation, Montreal

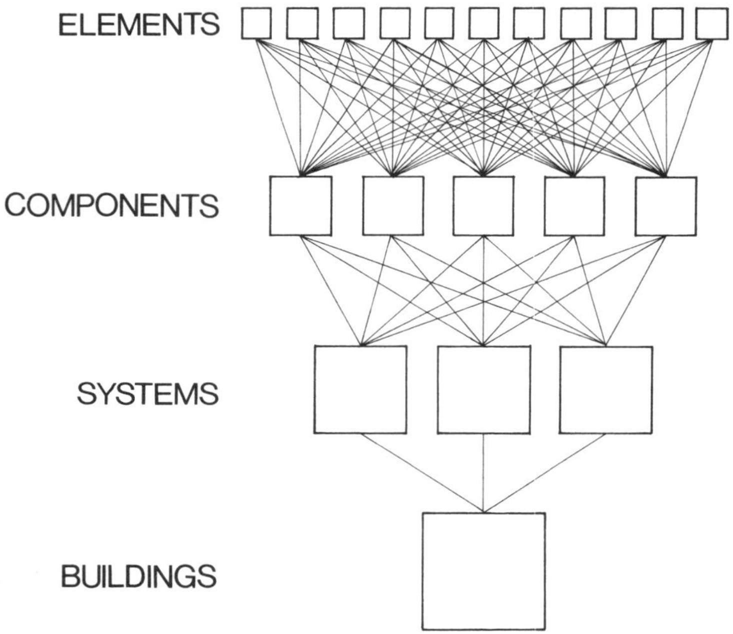

to the total building. These categories are: total building, building systems, building components, and building elements.

Total Building — The total building degree of prefabrication means that all the essential features of a building — the structure, the enclosure, the mechanical support services, and the finishes — are combined into the final product. The output is thus highly complicated and sophisticated. Questions of standardization and uniformity arise.

Building System — Prefabrication of a building system involves only one of the above-mentioned essential features of a building, such as structure. Here the assembly task is somewhat more specialized. As it is only one of several major factors in the final form of a building, the possible configurations are not as limited. The major problem that arises in this case is the matching-up, or interface, with the other systems of the building, which may also be prefabricated or conventionally constructed.

Building Components — Building components are the parts which make up a system. These parts can be assembled into a much higher number of possible configurations. However, this also means that they must be capable of a greater range of connections.

Building Elements — Building elements are the smallest pieces of a building which can reasonably be discussed in terms of prefabrication. Like the atom, they are the simplest level of reduction at which their unique character can still be identified. Building elements are the most flexible and versatile type of prefabricated parts with the largest range of possible uses and applications.

Systems building, or the systems approach to buildings, is a phrase that has enjoyed much popularity, but also much misunderstanding. As opposed to building systems which deal with a specific level of prefabrication, systems building has a much broader meaning. It considers a building as more than the physical pieces which make it up; it considers building as a process in which a set of interrelated and interdependent factors are combined and organized. The factors include not only the physical pieces which compose the building, but also such things as financing, management, labor, and marketing. Thus, the systems approach views the development and construction process holistically, as a whole. From this point of view, the solution to a building problem may not lie in a change in the way the parts are manufactured or put together, but may lie in a more rational, logical way in which it is organized or managed. A systems approach may not necessarily involve industrialized or prefabricated building elements as part of the solution.

Much more terminology has been generated to describe the various aspects of this field, and the reader is directed to the Glossary for further discussion.

* * *

Some characteristics of this change in the building process called industrialization are useful to investigate. The forms of industrialization can generally be broken down into four categories: rationalized building, standardization, new technology, and prefabrication.

'Scientific management,' value engineering, and other similar techniques can be applied to the construction process to attempt to increase productivity or efficiency. This is known as rationalized building. It does not necessarily require off-site production or new technology. What it does is to apply the knowledge of industrial production methods and techniques to the field of construction. By studying the organization, management and process of construction, better or more efficient ways of doing the same task are sought. Some examples of this approach are 'Fast-Track' and CPM (critical path method). It may also involve production line techniques either on-site or off-site.

Modular coordination and standardization are probably the earliest forms of industrialization to take place in building. Milled lumber and common brick are two of the basic traditional building materials that assumed uniform dimensions very early. This aspect of industrialization gains its importance in the efficiency of assembly. Standard-sized parts eliminate the need for costly and time-consuming custom fitting to the specific situation. Standard-sized parts can also be produced faster; there is less need to 'tool up' for differing production demands. Standardization is so prevalent as to have become unnoticed and commonplace. The variety of the built environment is sufficient example to demonstrate the degree of combinations and configurations possible. This raises the question of what is called 'open' and 'closed' systems. Depending upon how these standardized parts match up with other parts necessary for the complete building will determine their usefulness. If the parts only match up with similar parts produced by the same manufacturer, then it is a 'closed' system. But if the standardized part can be combined with a wider range of other parts, then it is an 'open' system.

Industrialization can also take place by the introduction of new technology and equipment. Machines can reduce the amount of time and labor required to perform a specific task. Power-driven nail

Automated truss nailing machine.
"GANG-NAIL" is the registered Trademark of Automated Buildings Components, Inc.

guns and 'flying' concrete formwork are some examples of how technological advancement can speed up the construction process.

Prefabrication is the most commonly perceived form of industrialization since it involves a change in how materials are received and installed at the site. This form of industrialization achieves its economies by removing stages of the construction process from the job site to some off-site production facility. The factory allows

Prefabricated metal wall system being erected.
United States Steel Corporation

incorporation of rationalized processes, standardization, and new technologies. It has better control over some of the problems of the job site, such as climate, quality control, and unit costs. It can utilize production line, mass production techniques. It is not without its problems and limitations, however, and depending upon the scale of the prefabricated element, it can have a varying impact on the construction process.

* * *

In whatever form industrialization takes, it does have an impact on the construction process. The time of construction can be shortened, whether due to improved scheduling, use of standardized parts, or use of prefabricated subassemblies. The quality of construction can be improved, especially by the use of prefabricated elements. This is due to the systematic assembly line process in which standardized, identical units are mass-produced. Cost is another factor which can be favorably improved. The cost reduction can be based on the actual cost of materials — a unit prefabricated by relatively unskilled labor will be less expensive than that constructed on-site by a skilled laborer; there is less likelihood of delays due to inclement weather since the prefabricated elements are constructed in controlled indoor conditions and simply installed at the site; there is also a reduction in the construction time. Industrialization, and especially prefabrication, can also lead to more varied designs since there is a greater variety of options from which to make a selection.

However, there is a reverse side to this coin. It does not follow that industrialization must produce these effects. As with all endeavors, the test lies in the execution. Inaccurately conceived and poorly executed attempts can be more disastrous than conventional construction. In addition, industrialization also brings with it its own set of peculiar problems. For mass production efficiencies to be realized there has to be a sufficiently high quantity of units produced. Some prefabrication processes may require a large capital investment for initial set-up. The transportation cost of moving the element from the production facility to the site may be substantial. There may be labor problems involving the learning of new skills or outright opposition by trade unions. There may be difficulties in obtaining approval by local code officials. And there simply may be reluctance to general acceptance.

There are many myths which have arisen about the industrialization of building, and some are seductive. Compared to other industries, the construction industry may appear to use outmoded, traditional methods and techniques. Industrialization is hailed as the way to 'catch up.' But this simplification fails to recognize some of the unique characteristics of building — its response to environmental, functional and behavioral needs.

Another myth is that the construction industry has been unable to meet the demands for building, specifically housing. Here industrialization is being viewed as the 'breakthrough' to solve this problem. But the issues are more complex than merely physical numbers. Economic and social issues among others are involved. Production of the objects is not the whole solution to the problem.

In spite of the shortcomings and false expectations about industrialization it does help in the solution of some of the problems. Especially at the small scale, industrialization has a valuable contribution to make to the betterment of the physical environment. In many ways prefabricated elements are an integral part of present-day construction practices. They have 'invisibly' been absorbed due to their beneficial aspects. Being small, they have been incorporated without causing a radical shift in the overall process or form. They have become part of the repertoire of building; they are the 'kit of parts' at the disposal of the designer. Acceptance by the users has been favorable. Perhaps the only problem is that in their 'invisibility' some of their potential has been overlooked; thus, the need is to make them 'visible.'

* * *

Wood modular unit being lifted into place.
Tom Mitchell, Jr., Photography

The magician's skill lies in his ability to make the visible invisible. In a sense the task of this book may be considered to make the invisible visible. Industrialized building elements are available and are possible building solutions — be they large or small scale. This handbook attempts to acquaint the user with this potential facet of building. Basic information about what the building elements are, their general characteristics and forms, is presented here. The intent is to demonstrate the range of possibilities available, to allow a larger data bank from which to derive solutions.

The book, therefore, is organized in a generic fashion. It describes elements by type rather than specifics. The information is intended to convey the logic and rationale of the approach rather than the details. To convey a sense of how they work and how they might be applied, general guidelines such as average sizes, typical forms, and typical details are provided. They are not intended to be considered as the only sizes or forms available, nor to represent the actual description of a building element produced by a specific manufacturer. The graphic, visual information is broken down and grouped into four main categories: Buildings; Systems; Components; and Elements. A description of these categories was provided earlier. In some cases, a specific type of part can be categorized in varying ways, generally due to its specific application. For such conditions, in this book, they have been listed in the division which corresponds to the way in which they are most commonly used, and reference made to their potential use in another classification.

Specifically, it is felt that the way to use this book is as an overview of what the options are. It can serve to assist in a preliminary selection of a specific type of building solution. After this point more detailed and specific information will be necessary. This information is not provided, nor could it be, in a book such as this. The common sources for this information are Sweet's File of manufacturer's information and the specific manufacturers themselves. Manufacturer's product representatives can be consulted to obtain the exact information about their product. Always consult specific manufacturer's data in a case of actual usage. The variation, adaptation, or combination of any general element described in this book comes into play at this level. The individual and the way that he uses this information are the critical factors in determining the usefulness of any of the variety of industrialized building elements.

It would be impossible to show every possible industrialized building element, every modification and improvement made by the individual manufacturer. The context and the framework are provided here. The specific form, characteristics, configurations, and availability will vary according to manufacturer and locale. The specific information changes over time and quickly outdates its publication. However, the basic rationale and logic underlying the specific form remain relatively constant over a longer period of time and that is what is presented here.

BUILDINGS

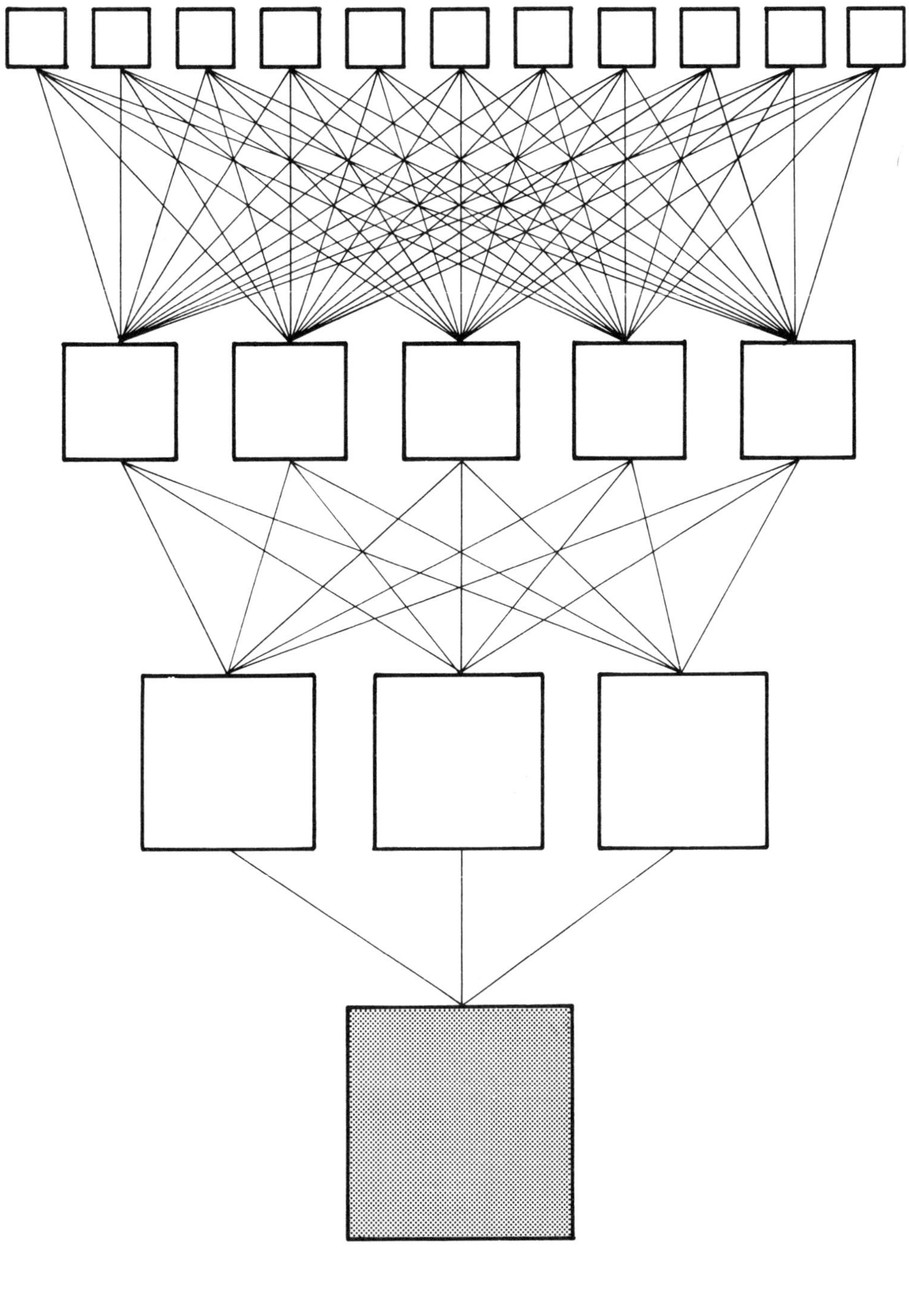

METAL

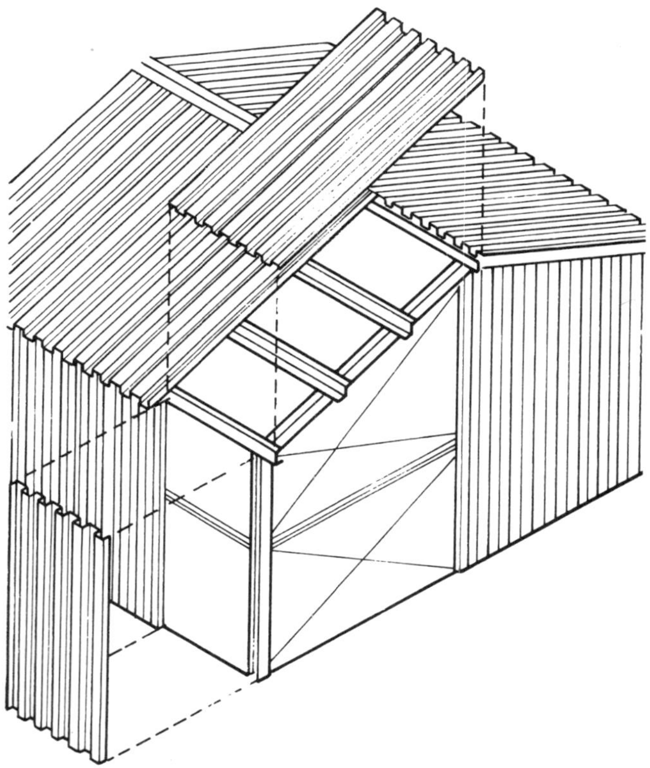

Metal buildings provide a total prefabricated building package. They consist of a complete set of preengineered, interdependent components, factory fabricated and ready for site assembly into a structurally sound, functional and economical building.

The structure consists of a steel frame which may be simple post and beam, tapered beam, mill bents, space frames, or truss. This allows for design flexibility in terms of size and configuration. Clear spans of 40 to 80 feet are possible. The average bay spacing is 20 to 35 feet, and heights range from 12 to 28 feet.

The enclosure system is generally ribbed metal siding and roofing. This siding is available in a wide range of profiles and colors. It can be insulated after installation or factory insulated. A large choice of accessories, such as double glass doors, fixed and operable window units, are integrated to be compatible.

The interior can be subdivided by demountable partition systems detailed to assure integration with the structural and enclosure systems. Integrated ceiling systems are also available which provide a combined lighting/ceiling subsystem. This subsystem provides illumination, acoustic control, partition support, and air supply and air return via the ceiling plenum.

The advantages of prefabricated metal buildings are: faster occupancy, generally two-thirds the time of conventional building; lower initial cost, as much as 25 percent less than conventional construction; reduced maintenance and lower operating cost; ease of future expansion; and flexibility and compatibility.

Many manufacturers, generally through local authorized builders, can provide a total turnkey construction service. The general market involves commercial and community use facilities, as well as manufacturing buildings.

types

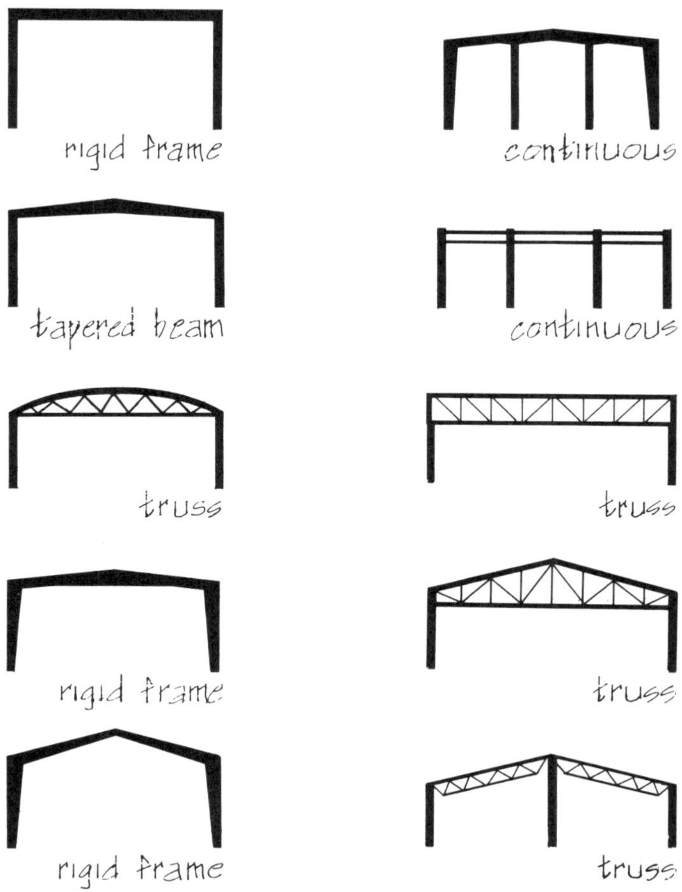

tapered beam

clear span widths	20', 24', 30', 36', 40', 50', 60'
wall heights	10', 12', 14', 16', 20', 26'
live loads	12, 20, 30, 40 psf
bay spacing	16', 20', 24', 25'

rigid frame

clear span widths	40', 50', 60', 70', 80', 100', 120', 140'
wall heights	10', 12', 14', 16', 20', 26'
live loads	12, 20, 30, 40 psf
bay spacing	16', 20', 24', 25'

continuous/modular

total widths	60', 70', 80', 100', 120', 140', 150', 160', 180', 200', 240', 300'
module widths	30', 40', 50', 60'
wall heights	10', 12', 14', 16', 20', 26'
live loads	12, 20, 30, 40 psf
bay spacing	20', 24', 25', 30'

truss

clear span widths	30', 40', 50', 60', 70', 80', 90', 100'
wall heights	10', 12', 14', 16', 18', 20', 22', 24'
live loads	12, 20, 30, 40 psf
bay spacing	20', 24', 25'

corner detail

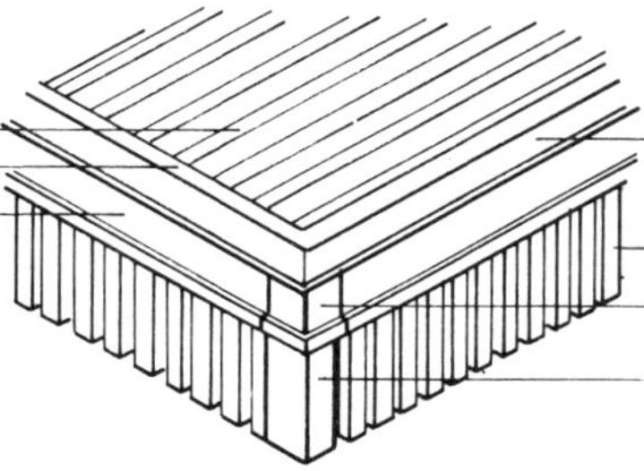

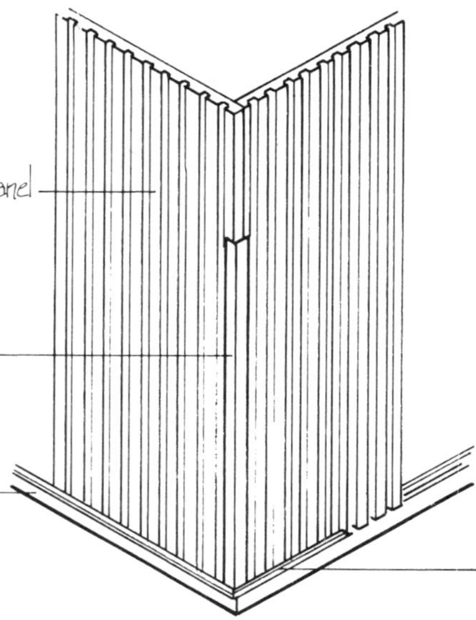

panel types

insulated

uninsulated

insulated panel

allowable wind pressure (psf)											
panel gauge	Fy	span (ft)									
		5	6	7	8	9	10	11	12	13	14
26	33	63	49	39	32	27	22	19	16	14	12

uninsulated panel

allowable wind pressure (psf)					
panel gauge	Fy	span (ft)			
		4	5	6	7
26	33	50	32	22	16

connections

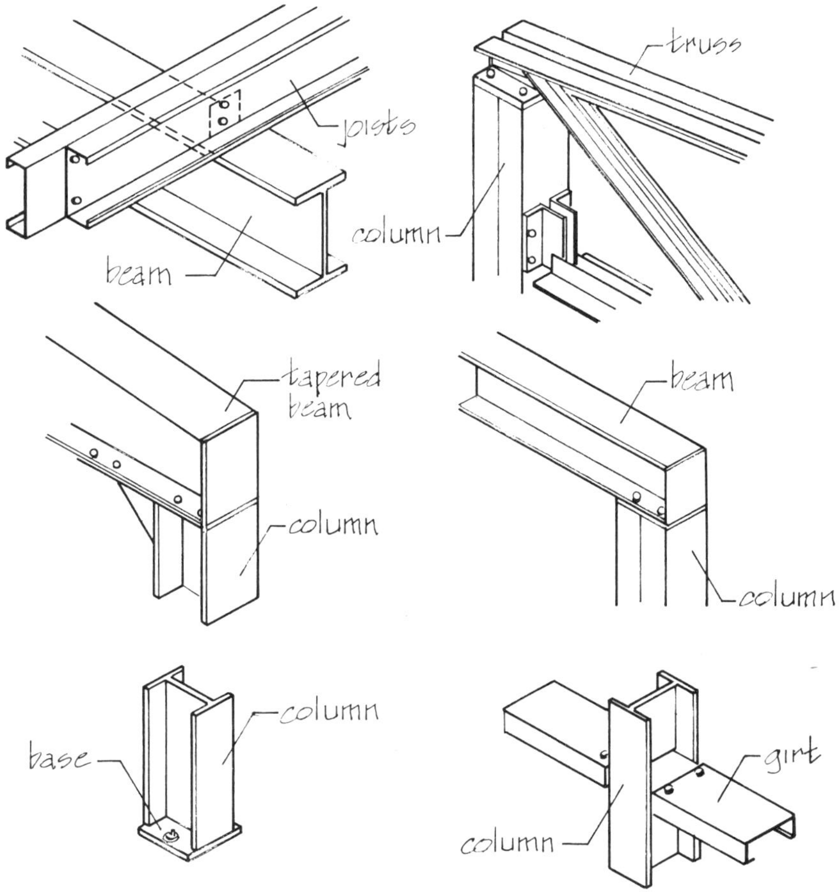

details

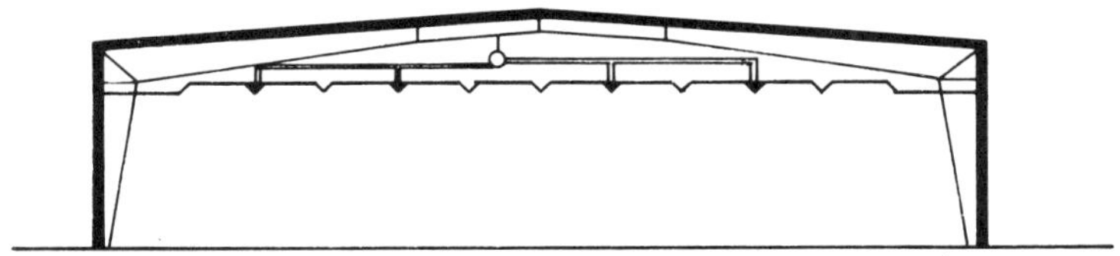
integrated ceiling

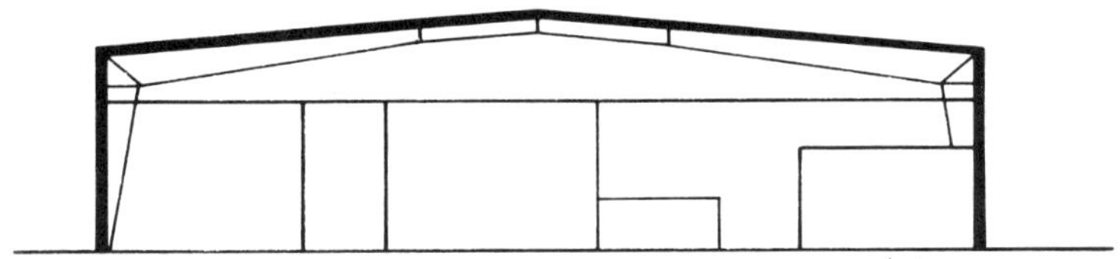
interior partitions

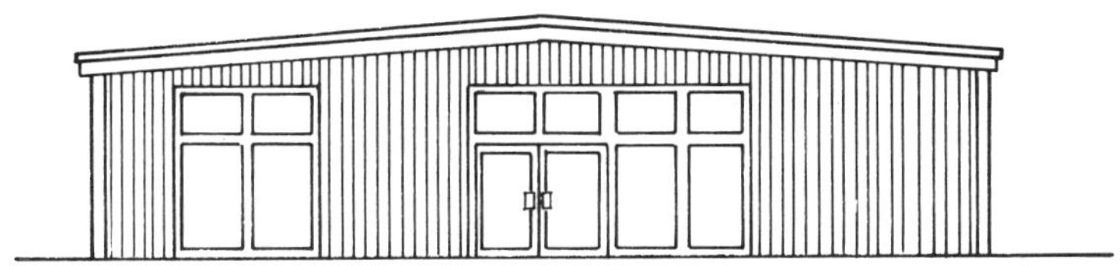
openings

CONCRETE MODULAR

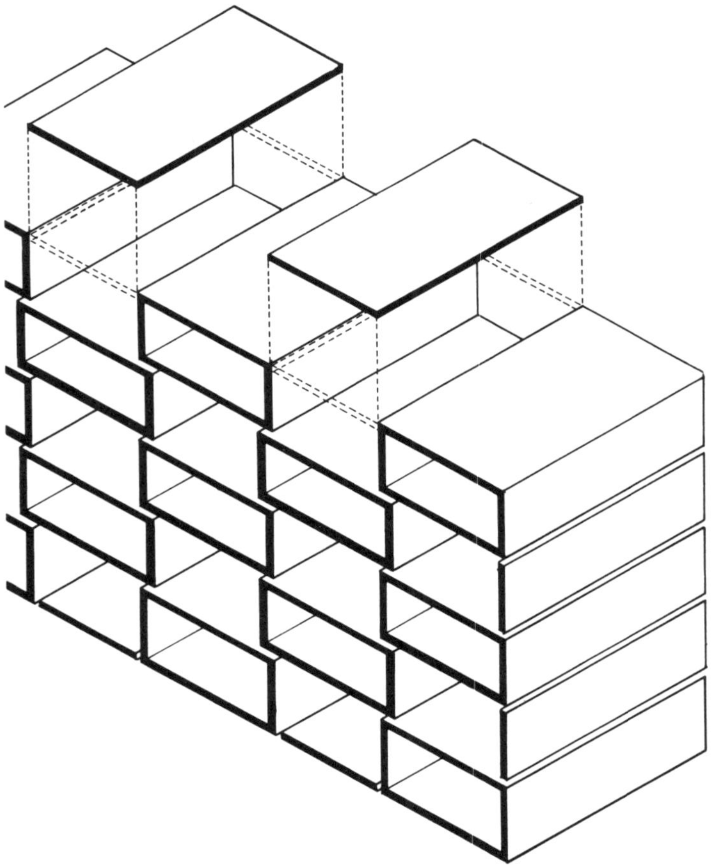

Concrete modular buildings consist of factory produced, three-dimensional, volumetric modules which produce complete buildings when combined with other similar units and related building elements such as stairs and elevator shafts. The modules are monolithically cast, box-like concrete units of variable size, usually one or two rooms large.

The modules can be combined and stacked in a variety of configurations, producing a wide range of building forms from townhouse to high-rise. The combinations possible allow for extremely flexible floor plans and building form designs.

The concrete modules are produced and extensively finished in the production facility. There is quite a range of finishes available, both for exterior faces and interior floor, wall, and ceiling surfaces. As part of the factory production operation, mechanical and electrical services are integrated and cast in place and are simply interconnected at the job site. By the arrangement of the modules, vertical utility shafts can also be created.

Since the units are so complete as delivered to the job site, construction time can be greatly reduced, sometimes as much as 50 percent. Costs can also be reduced, in some cases as much as 25 percent.

Precast concrete modular buildings have found applications as residential, commercial, and educational facilities.

plan configurations

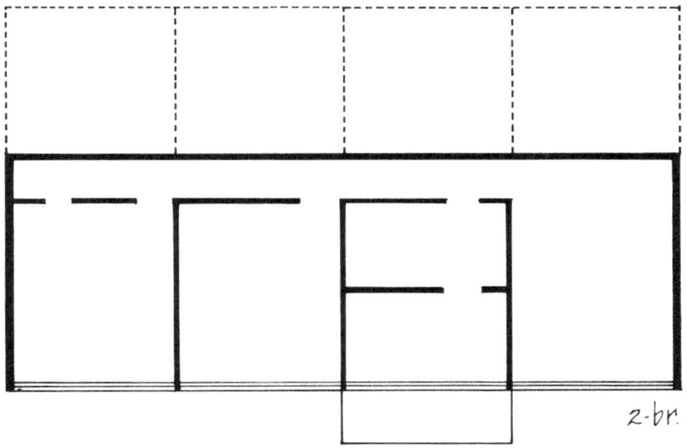

3 half modules

2-br.

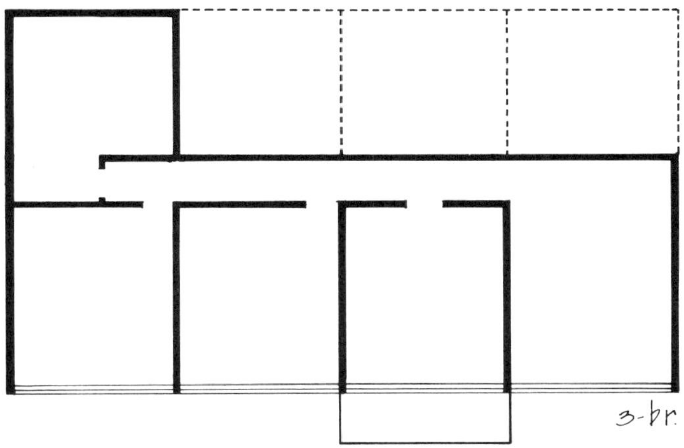

3 half modules 1 full module

3-br.

sizes

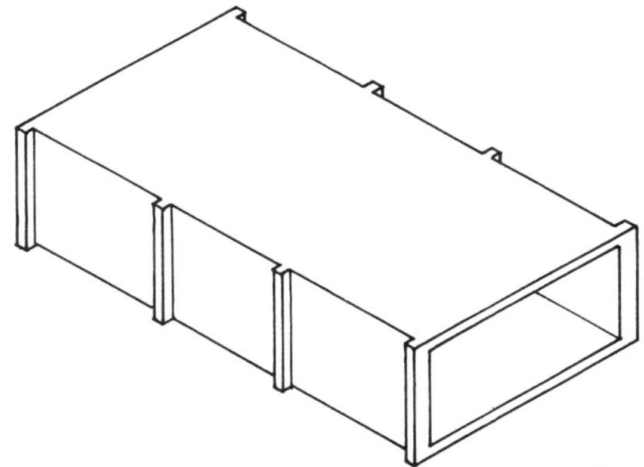

typical module

sizes

length	20 to 60'
width	up to 14'
weight	25 to 60 tons

configurations

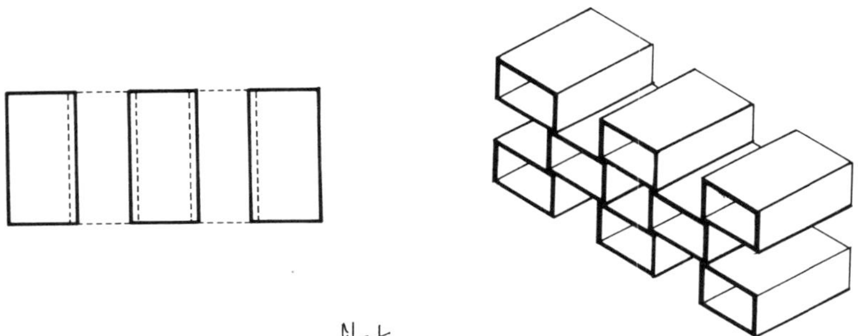

flat

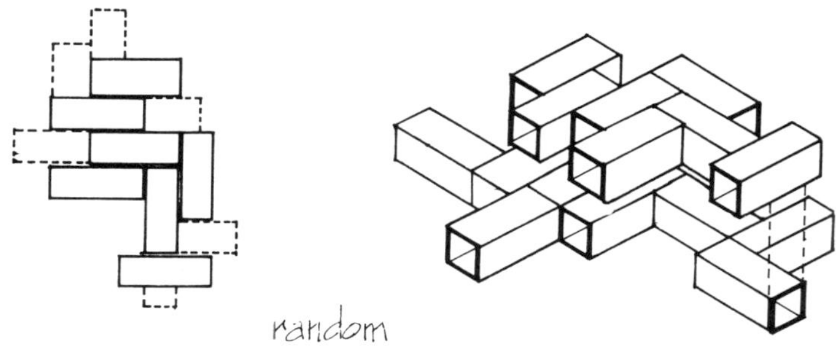

random

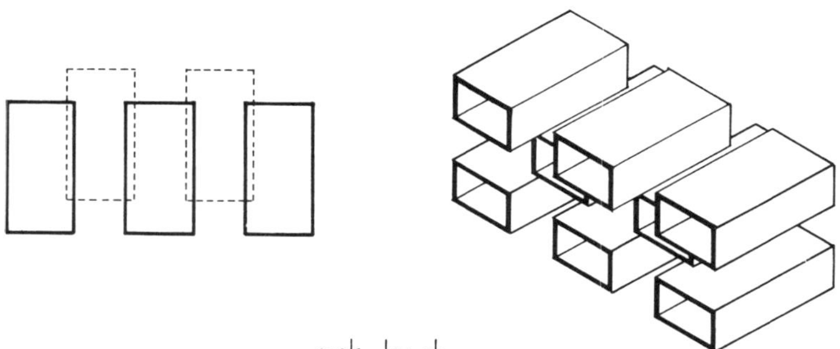

set-back

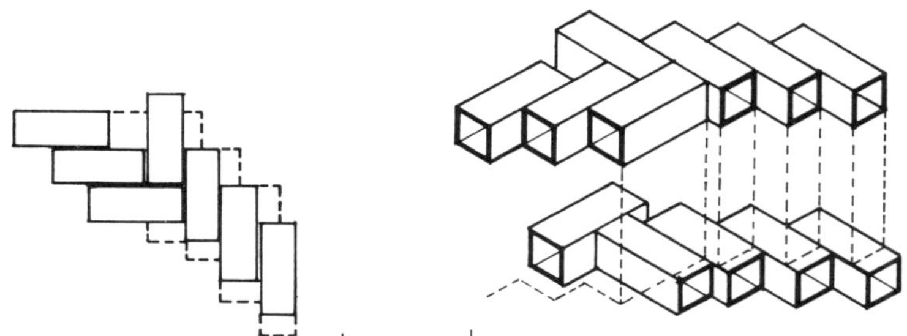

staggered

details

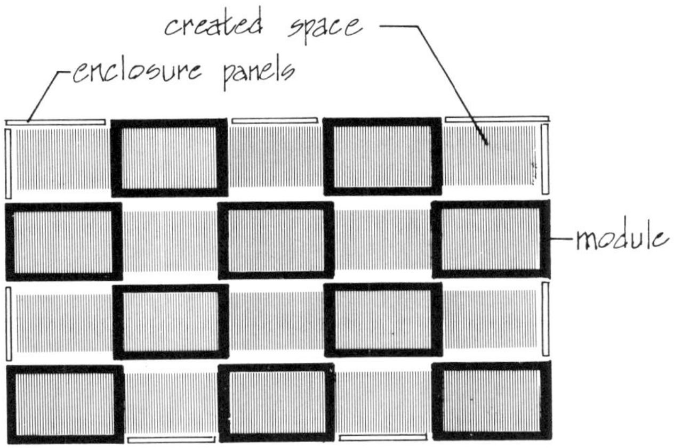

structural diagram

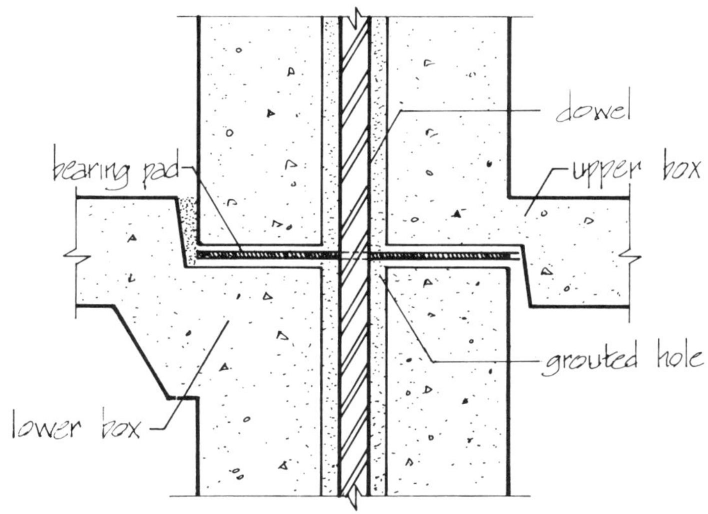

CONCRETE PANELIZED

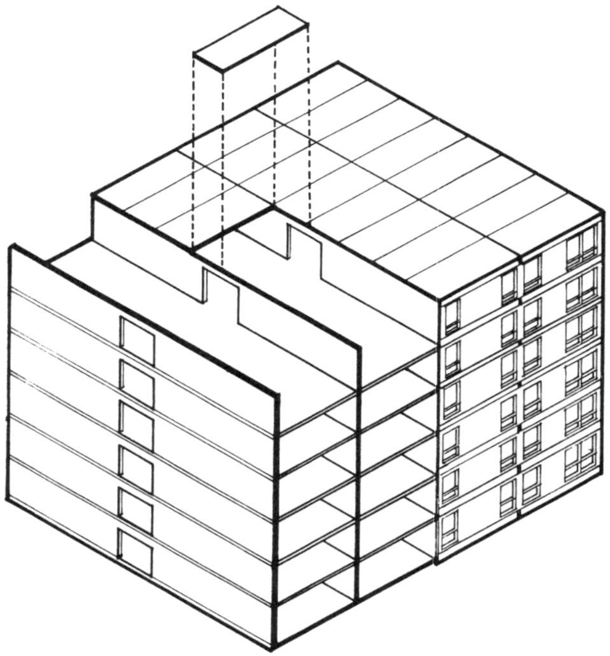

Panelized concrete buildings consist of floor and wall segments that are precast in a factory and assembled on-site. The system is basically composed of load-bearing wall panels, which may be interior or exterior, combined with floor and roof panels and non-load-bearing partitions to create an integral structure. The wall panels contain all necessary electrical, plumbing, and heating-ventilating services, which are installed during the precasting process. They also incorporate all the various openings required, such as door and window frames. Large mechanical areas, such as kitchens and bathrooms, may be separate units, complete with fixtures and accessories, which are manufactured in the factory and installed at the site.

The load-bearing wall elements are usually 8 feet in height and can be precast as single wall length sections or in shorter partial sections for ease of transport and erection. The floor panels can be large, room-sized slabs, or precast hollow-core planks.

A variety of options is possible for finishes depending upon the design. The exterior panels can be left as constructed, or treated. The interior wall surfaces are generally smooth finished, ready for application of interior treatment such as painting. The floor slabs are also smooth finished and can have carpeting, wood flooring, tile or plastic surfaces installed. The ceilings can be finished by application of paint, sprayed plaster, or acoustic tile, and dropped ceilings are also possible.

In addition to the general advantages of concrete construction, such as its weathering characteristics and good sound resistance and thermal capacity, panelized concrete buildings can reduce on-site labor to a minimum; it consists primarily of site preparation, erection, securing, and minimal finishing operations. Panelized construction can be applied to high-rise buildings as well as low-rise and townhouse type construction.

configurations

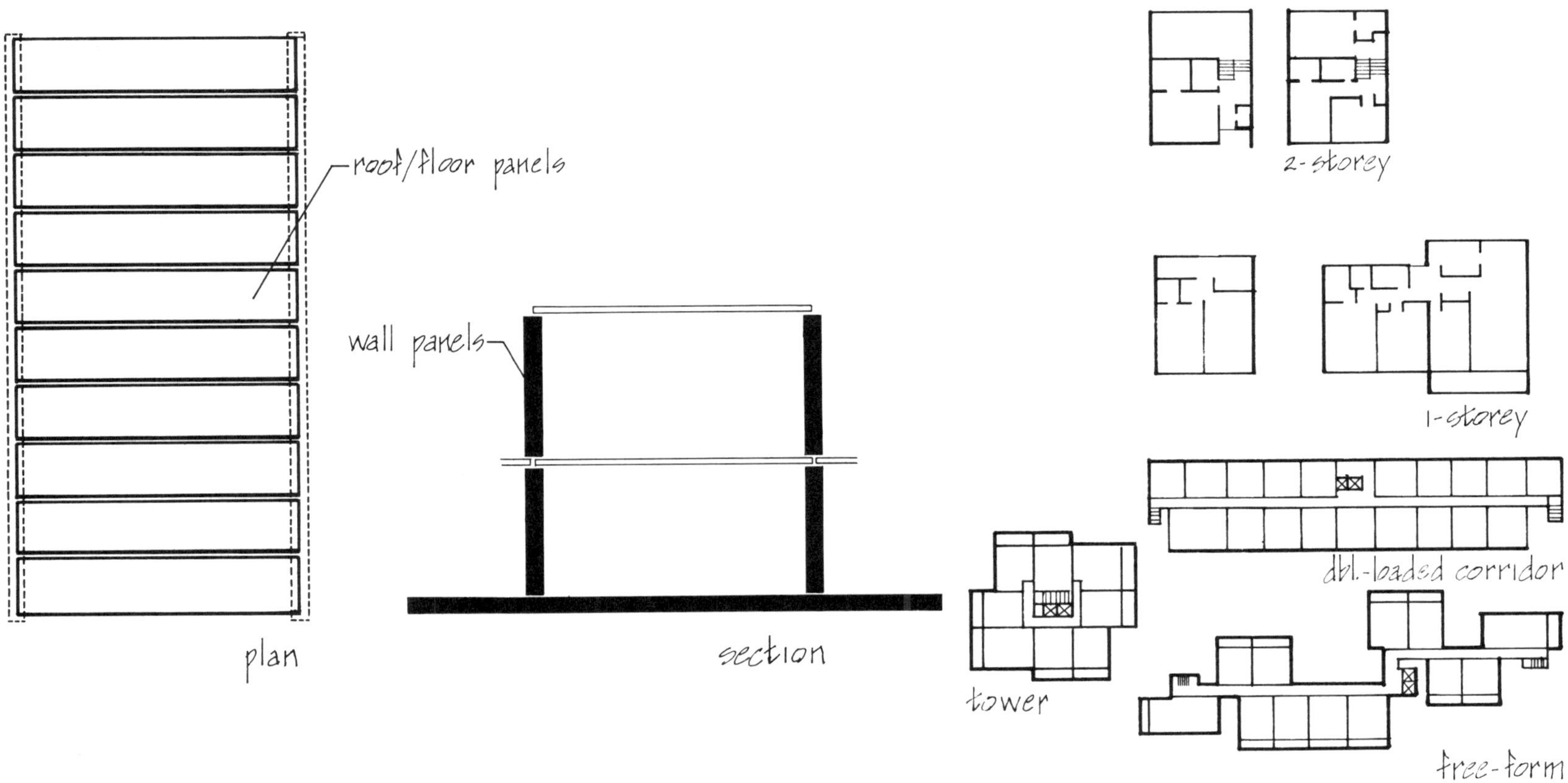

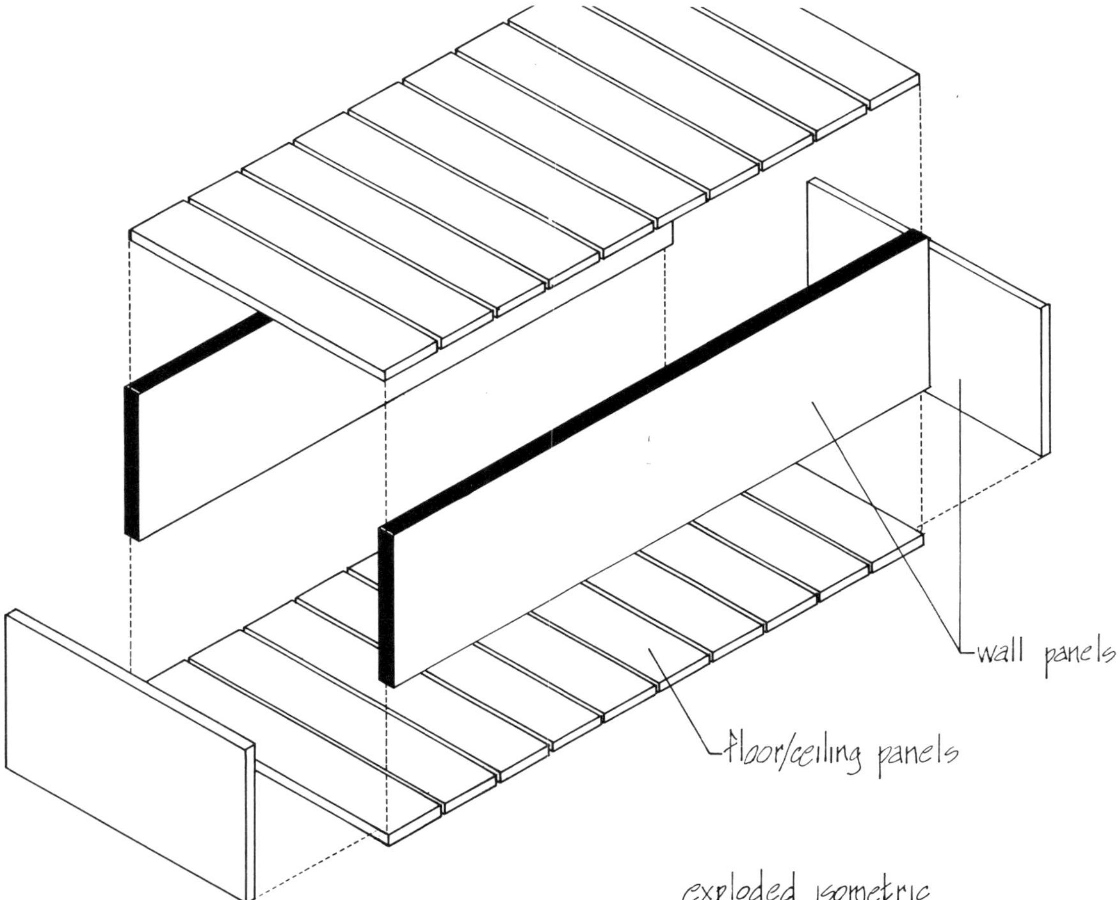

exploded isometric

typical sizes

floor panels

length	22' to 24'
width	3'-8" to 10'-8"
thickness	6" to 8"

wall panels

length	up to 40'-0"
thickness	8", 10"

details

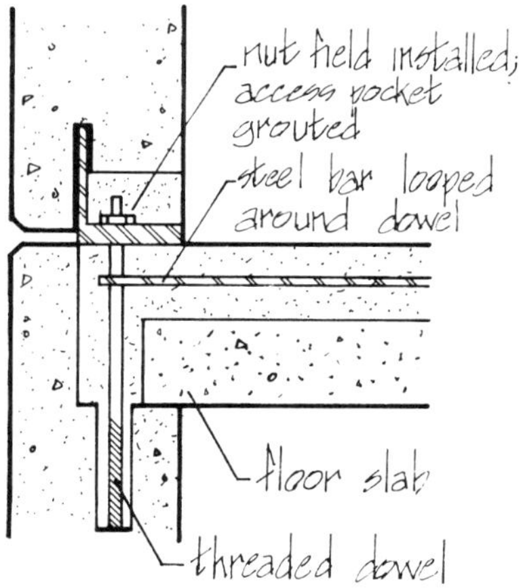

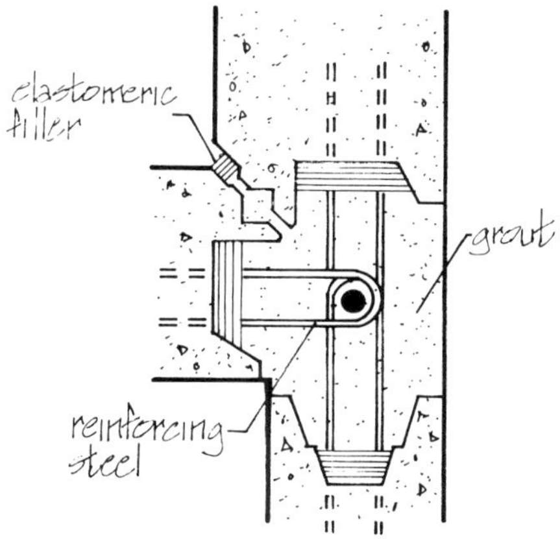

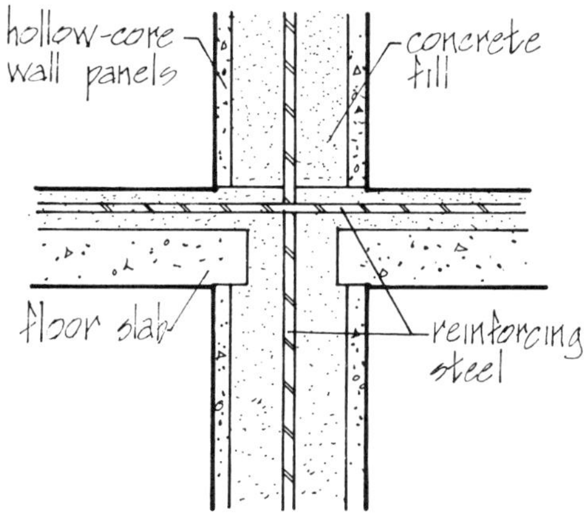

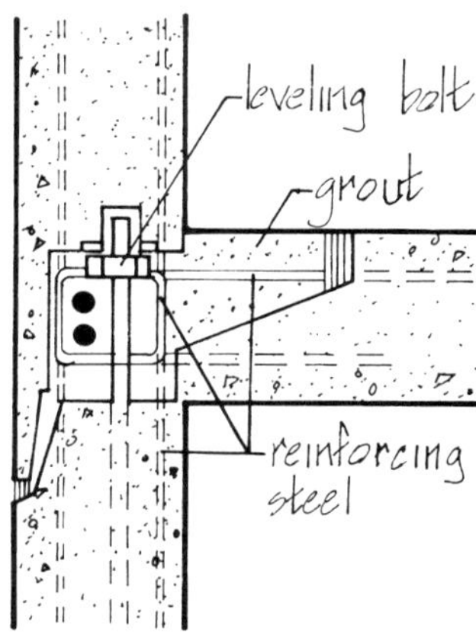

applications

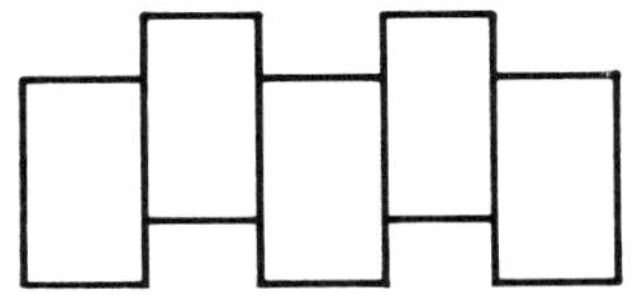

townhouse

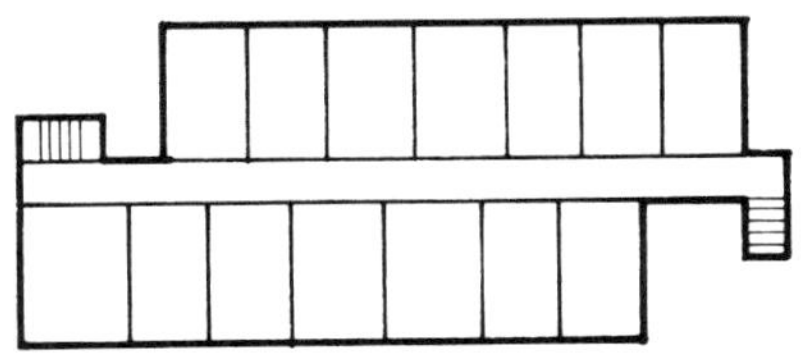

double-loaded corridor

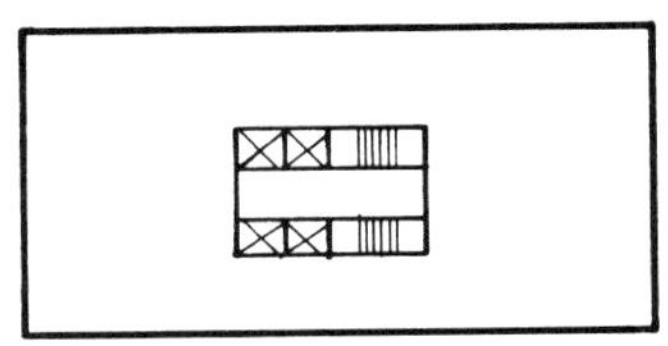

high-rise

WOOD MODULAR

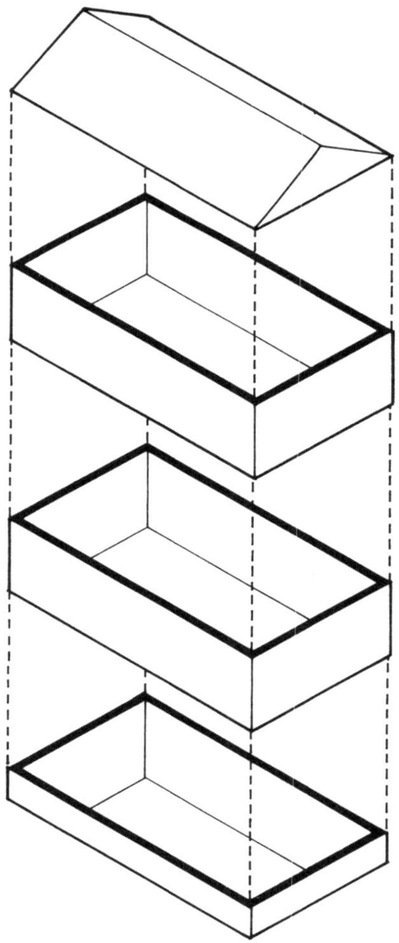

Modular wood buildings are composed of units, or modules, of relatively conventional wood frame construction, that are preassembled in a factory and combined at the job site. The units are three-dimensional volumes of standardized external dimensions, but capable of a variety of internal subdivisions. The assembly at the site consists of placing the units into position by crane and making the necessary connections.

Generally, the exterior of the modules is the exposed studs, allowing easier field connections and field application of exterior finishes. The units have the basic mechanical and electrical service lines installed in the factory, which are simply hooked up at the site. The interior surfaces may be left rough, partially finished, or totally finished in the factory, depending upon the project requirements.

Structurally, since they are of conventional wood framing construction, the maximum stacking capability is limited to three units high. Many configurations are possible, although they are generally used for standard townhouse, garden apartment, duplex, and single-family dwellings. They have also been used as motels and schools.

configurations

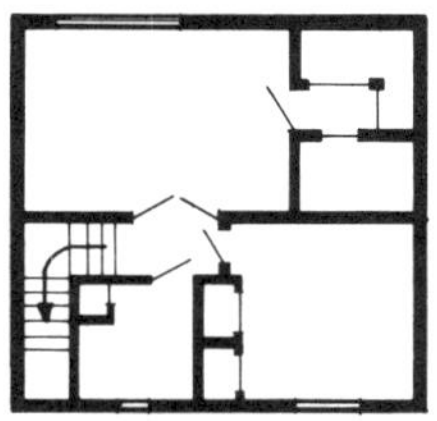

2 nd. fl.

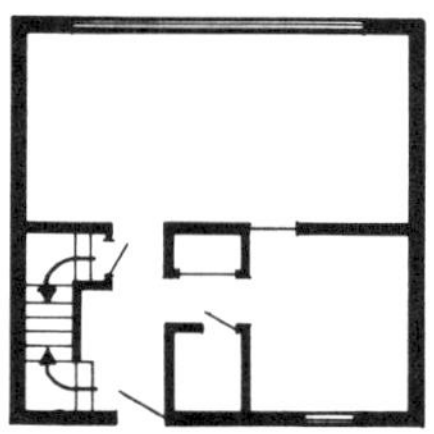

1 st. fl. 2-bedroom

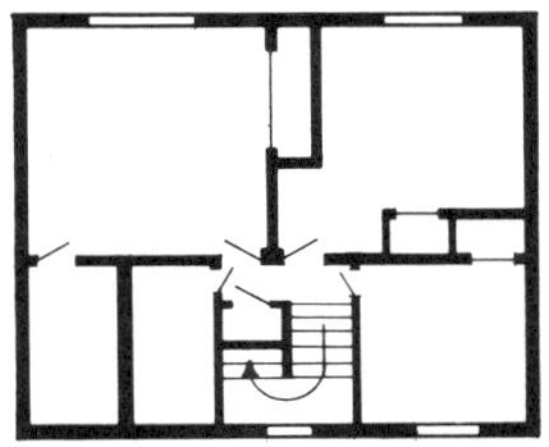

2nd. fl.

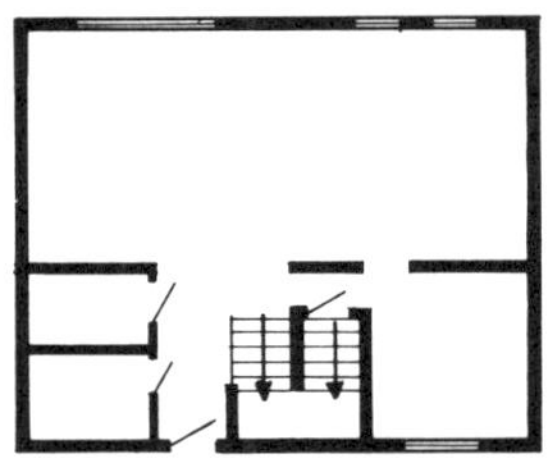

1st fl. 3-bedroom

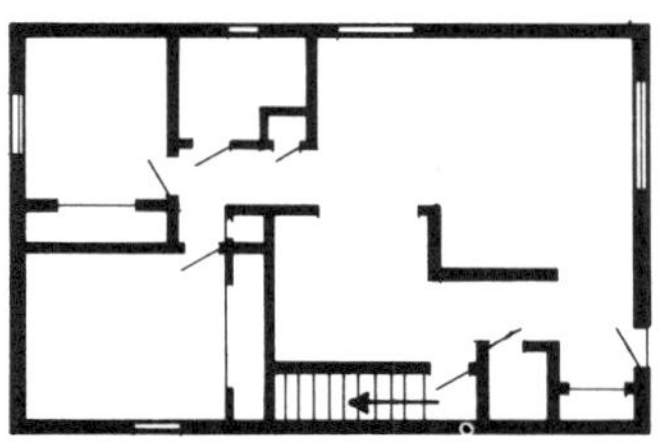

1-storey 2-bedroom

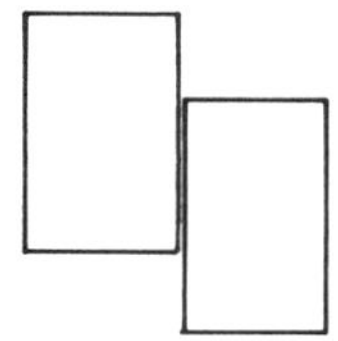

set-back

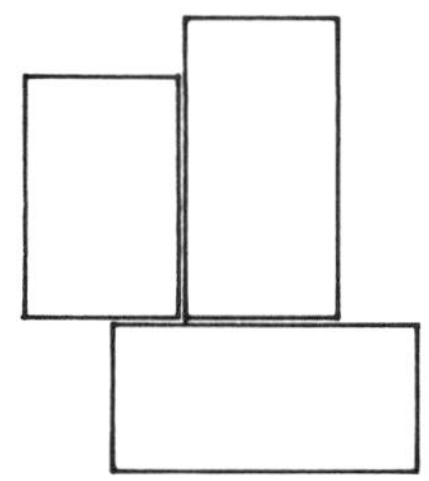

rotated

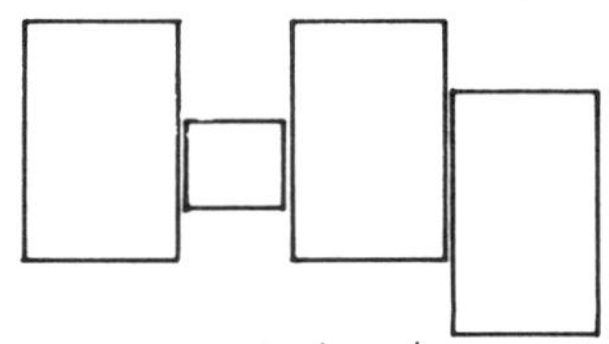

set-back with service core

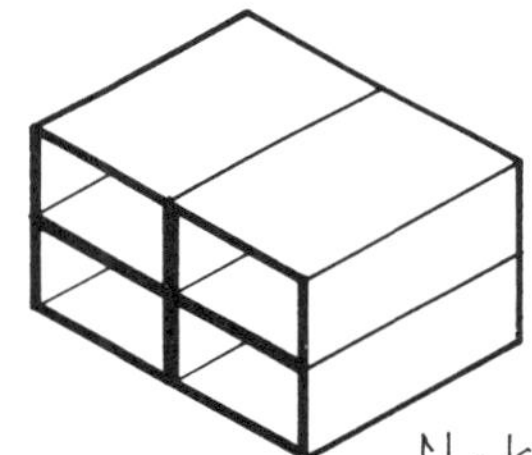

flat

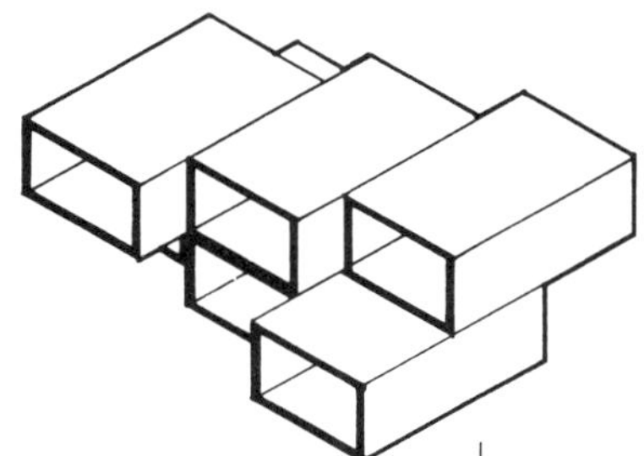

staggered

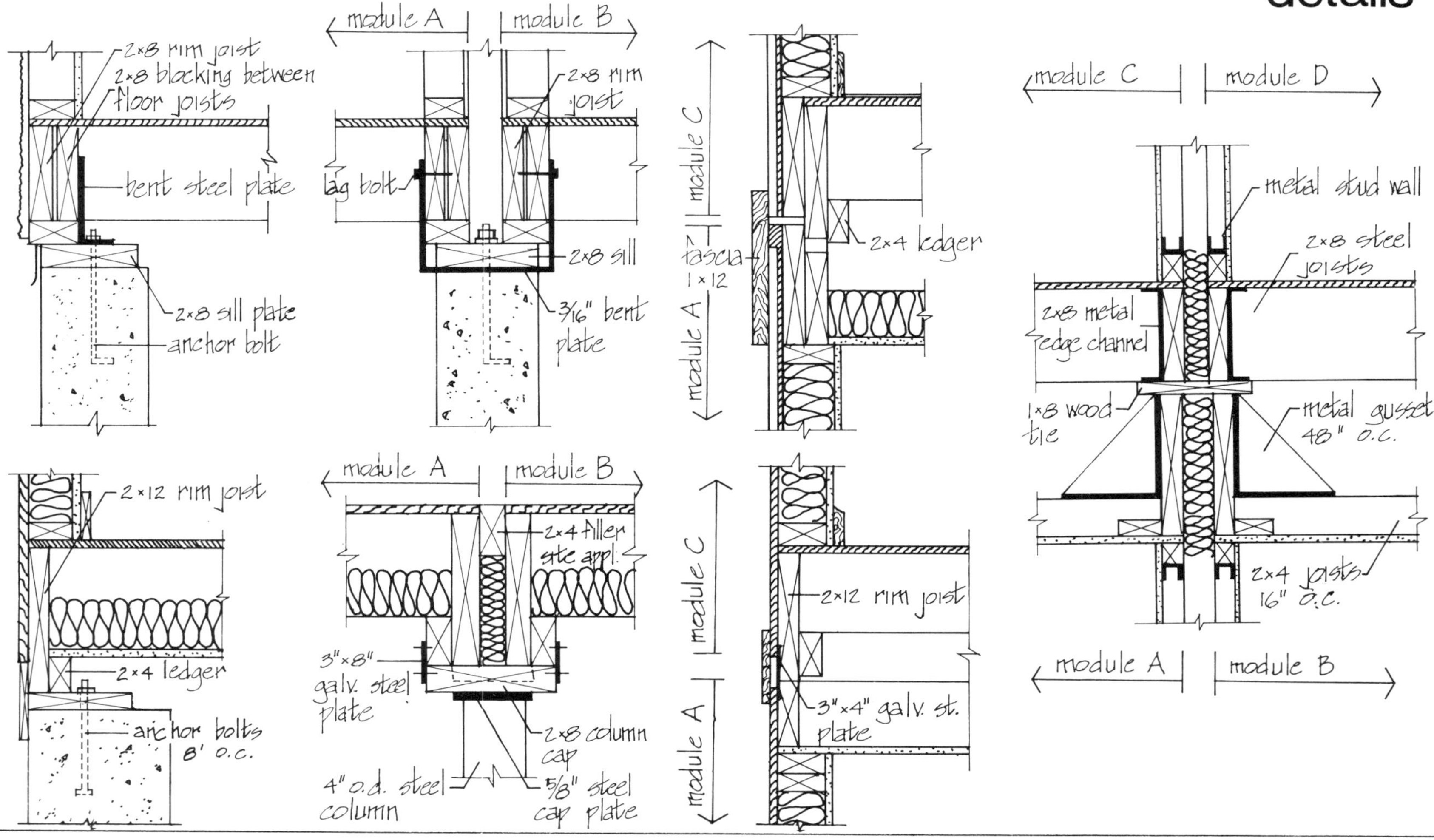

2×8 rim joist
2×8 blocking between floor joists
bent steel plate
2×8 sill plate
anchor bolt
module A
module B
2×8 rim joist
lag bolt
2×8 sill
3/16" bent plate
module C
module A
fascia 1×12
2×4 ledger
module C
module D
metal stud wall
2×8 steel joists
2×8 metal edge channel
1×8 wood tie
metal gusset 48" o.c.
2×4 joists 16" o.c.
module A
module B
2×12 rim joist
2×4 ledger
anchor bolts 8' o.c.
module A
module B
2×4 filler site appl.
3"×8" galv. steel plate
2×8 column cap
4" o.d. steel column
5/8" steel cap plate
module C
module A
2×12 rim joist
3"×4" galv. st. plate

applications

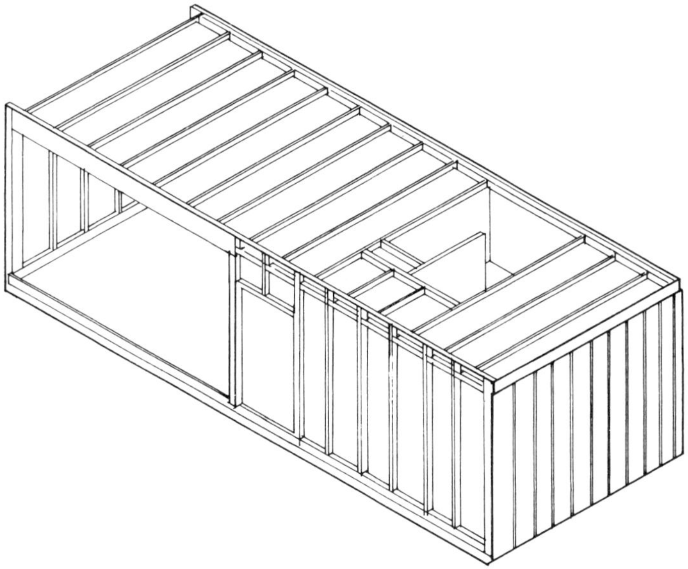

structural cut-away

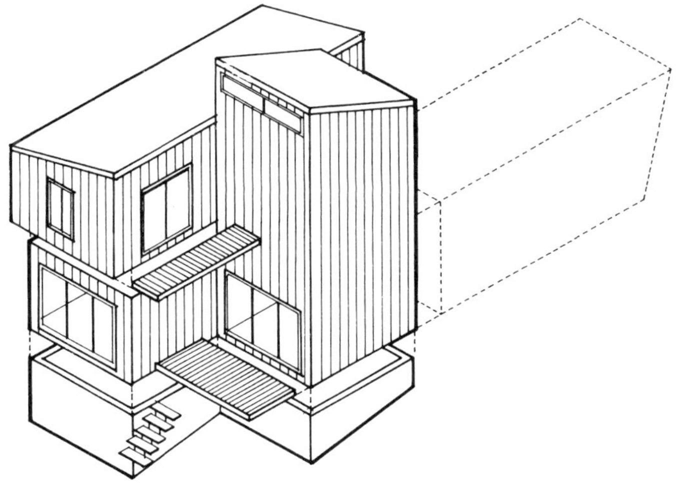

horizontal and vertical tilt-up modules

WOOD SECTIONAL

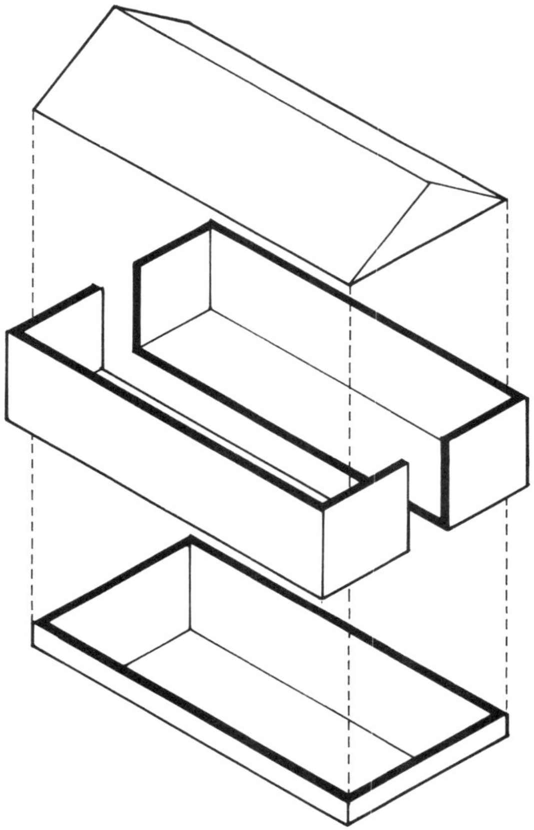

Sectional wood buildings are a sort of specialized case of modular wood building. They are almost exclusively considered in terms of residential construction. The basic form of sectionals is a long, narrow volumetric module, complete on three sides, which when joined with another similar unit produces a complete building.

The size of sectionals is generally limited by the maximum size permitted for highway transportation. The construction is of standard wood framing, although the floor may be stronger to handle the transportation and erection loads.

The units are almost totally finished in the factory, site work being limited mainly to foundation work and the actual connection between the two halves. A variety of plan configurations is possible, as is a choice of exterior and interior finishes.

plans

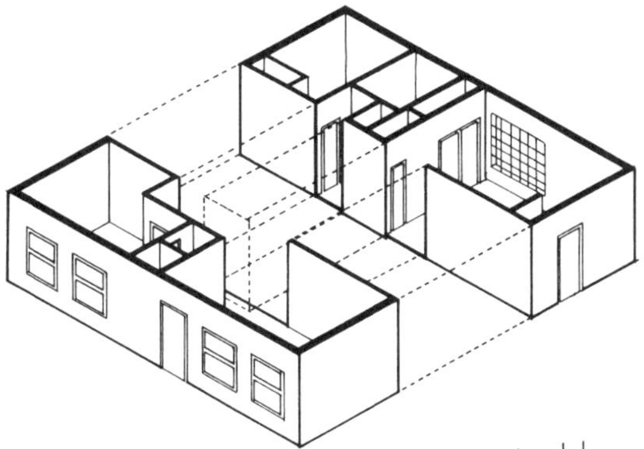

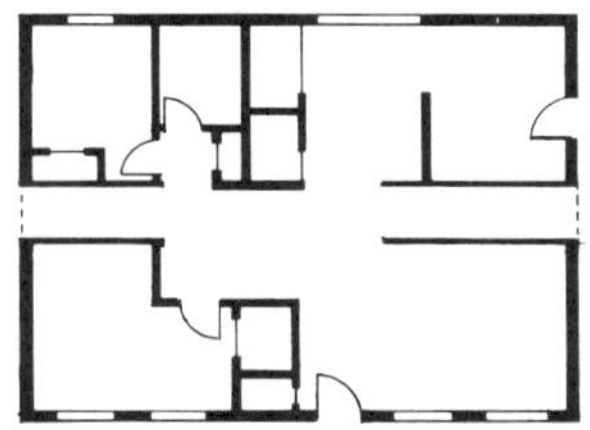

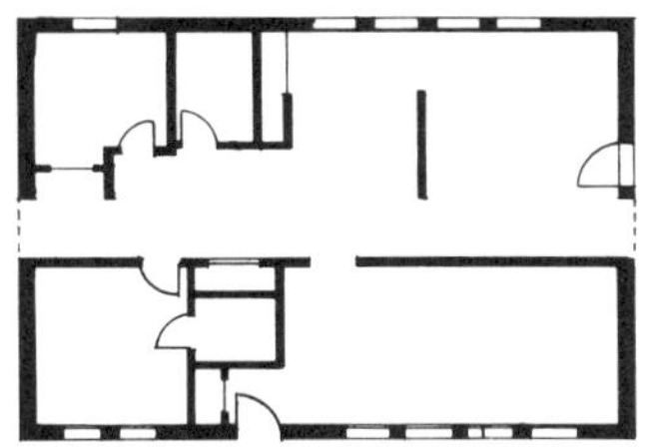

double wall

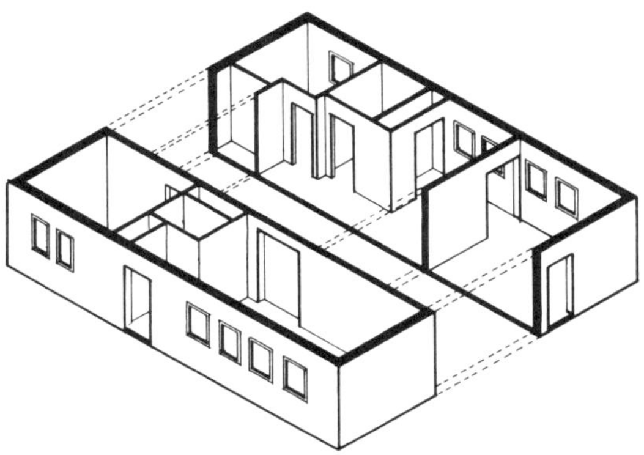

single wall

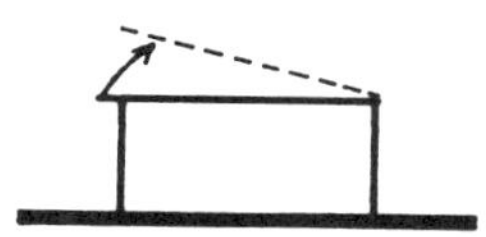

tilt-up roof

details

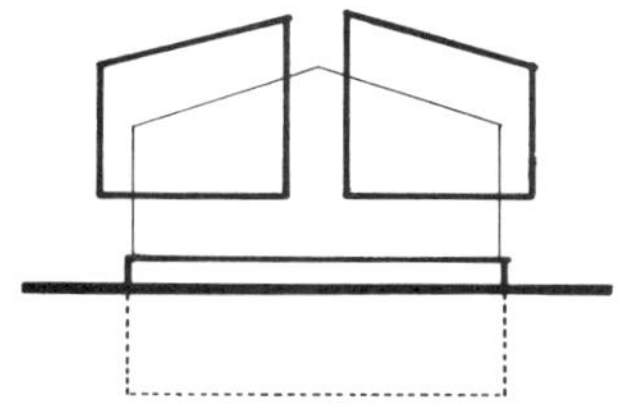

2-part

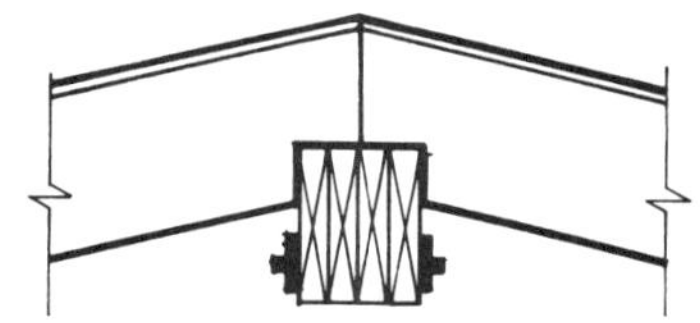

ridge beam

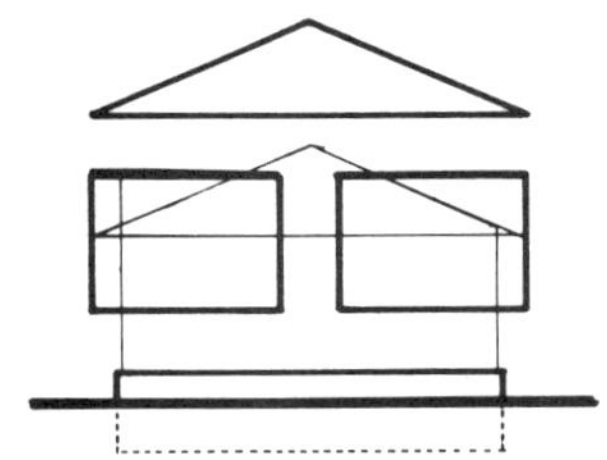

3-part

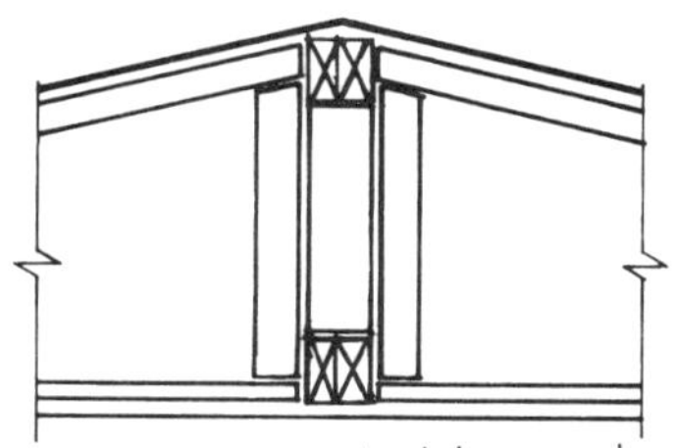

butt joint

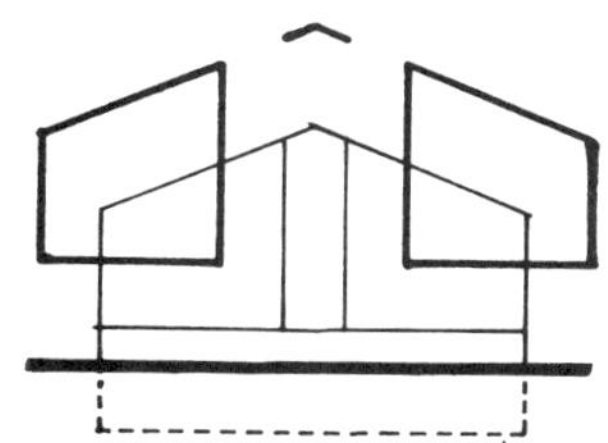

corridor

floor connection

WOOD PANELIZED

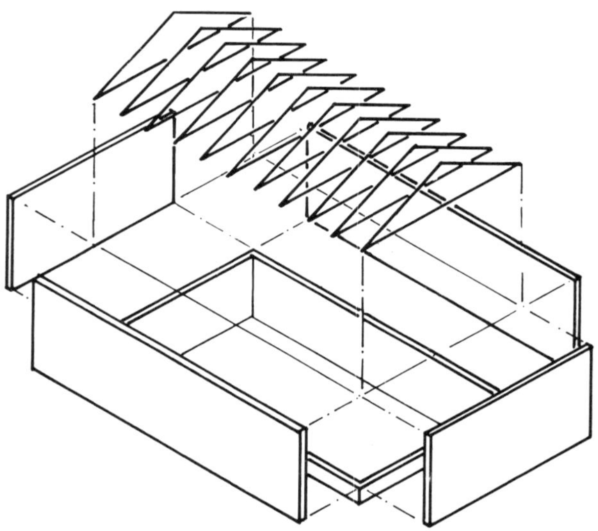

Panelized wood buildings are similar to panelized concrete buildings in that the component parts of floor, wall, and ceiling are factory produced into subassemblies that are then combined at the job site into the final building.

The structure can be either conventional wood framing or stress-skin panel design. The panels can range in degree of finish from simple studs and plates, to sheathing on one side, to completely finished on both sides.

The panels can be factory fabricated to meet the needs of the specific design, and thus can produce practically any form that can be made with conventional wood framing. Panelized wood buildings have found applications in small residential and commercial uses, although their primary application is to single-family houses.

plans

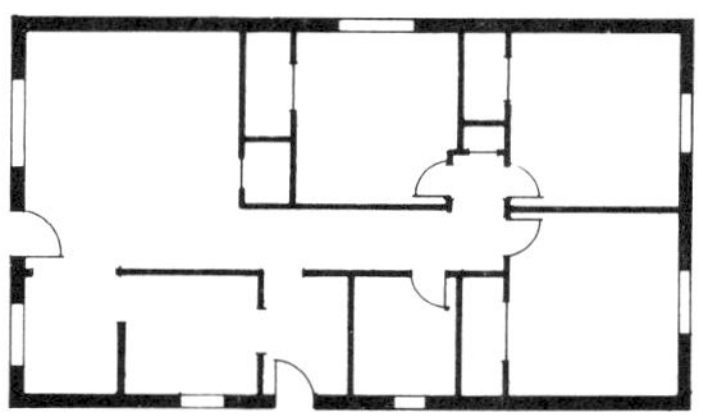

rectangle

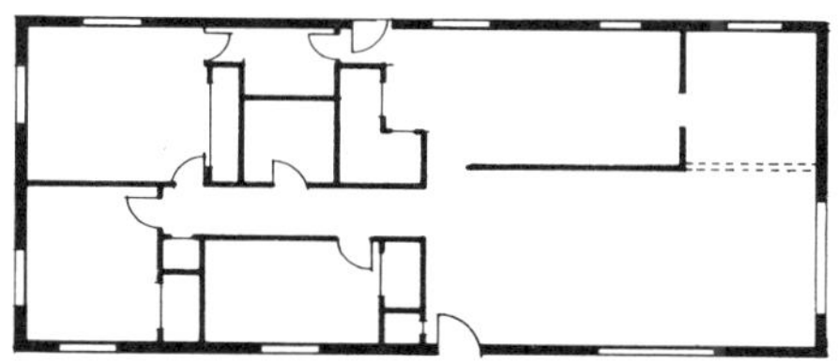

elongated rectangle

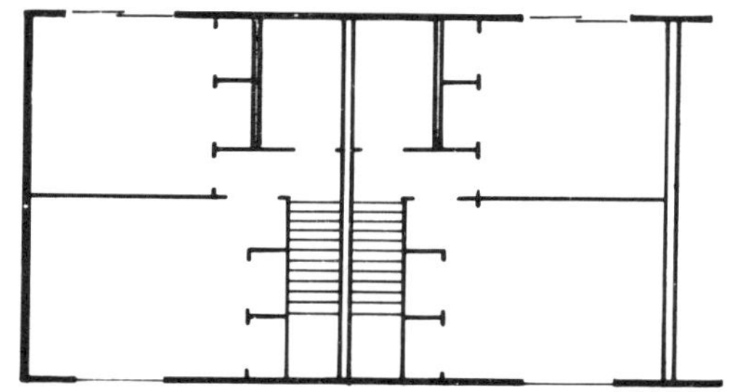

2 nd. floor

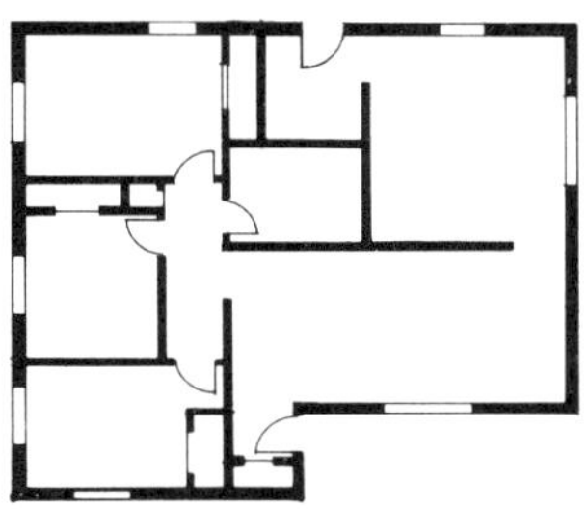

L - shape

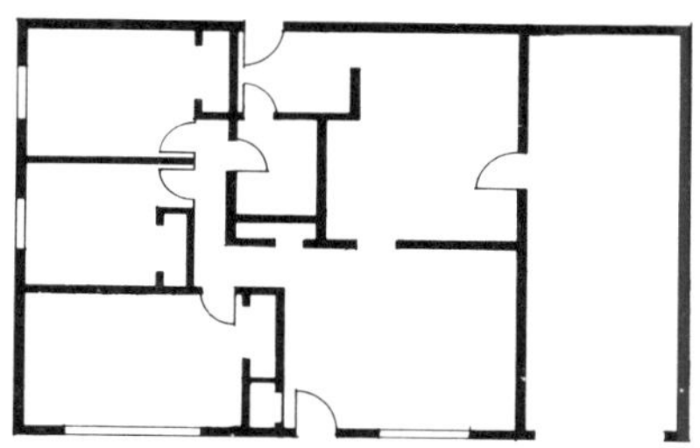

attached garage

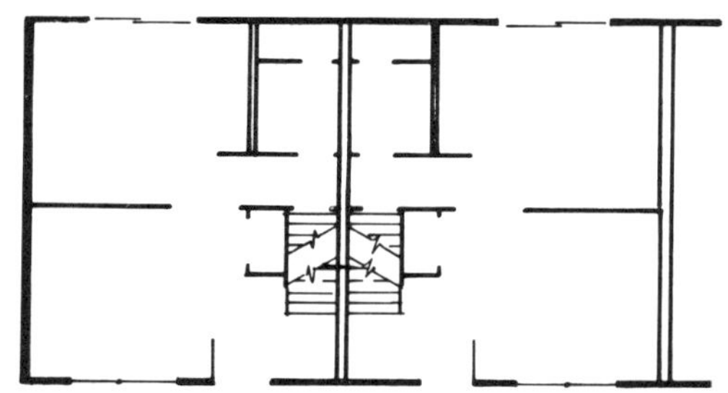

1 st. floor

garden apartment

assembly

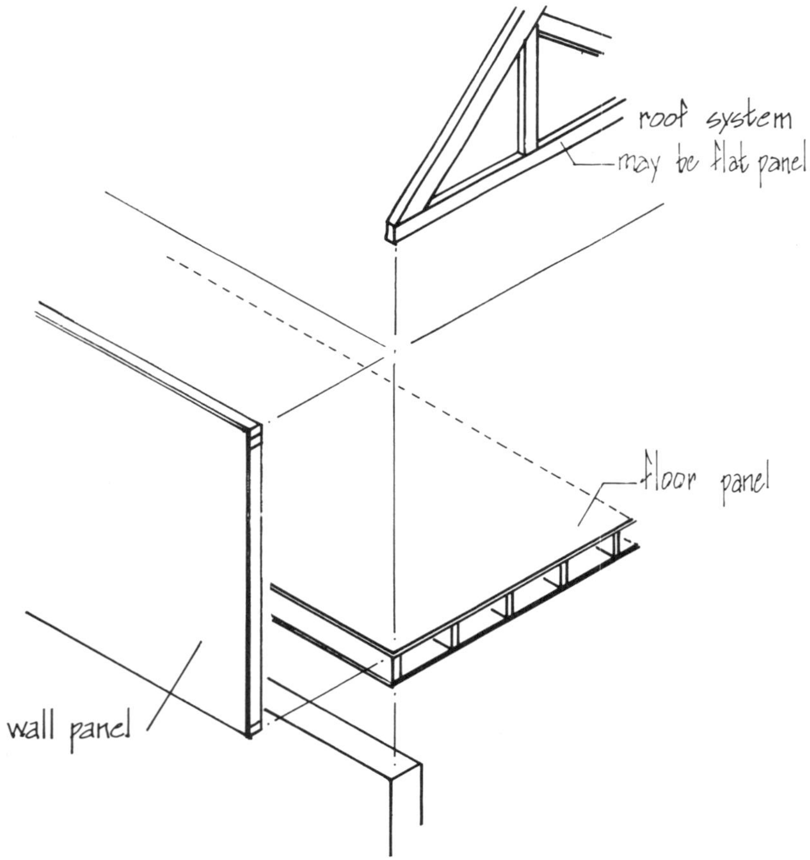
roof system
may be flat panel
floor panel
wall panel

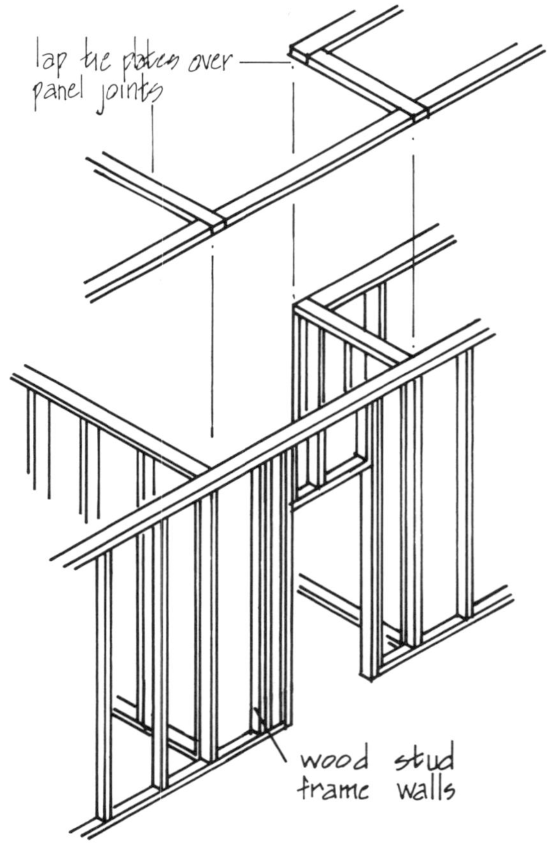
lap tie plates over panel joints
wood stud frame walls

details

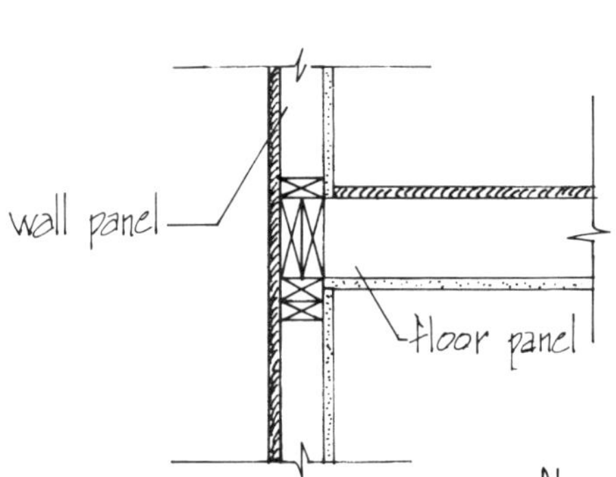

floor to floor

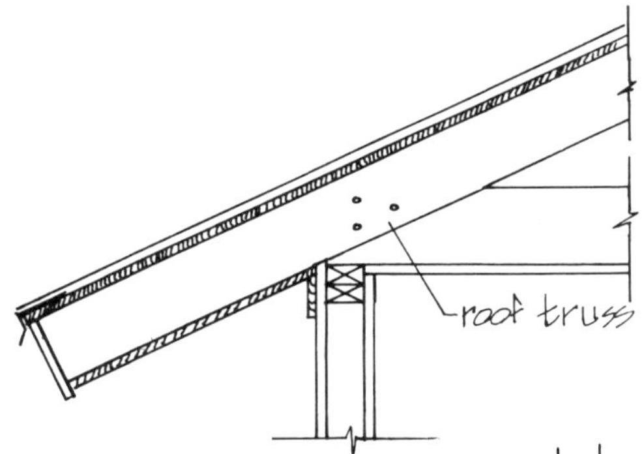

pitched roof

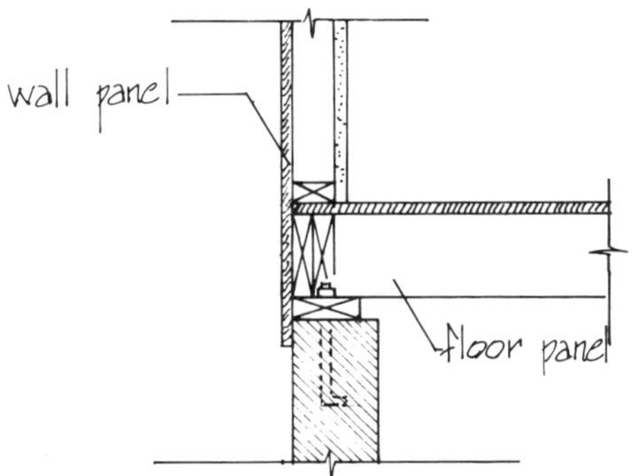

floor to foundation

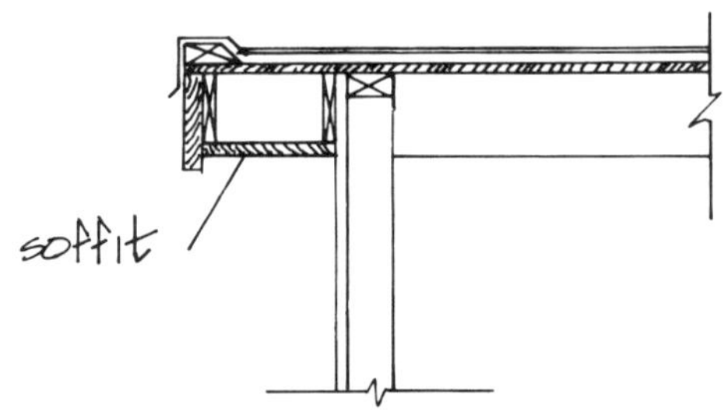

flat roof

WOOD PRE-CUT

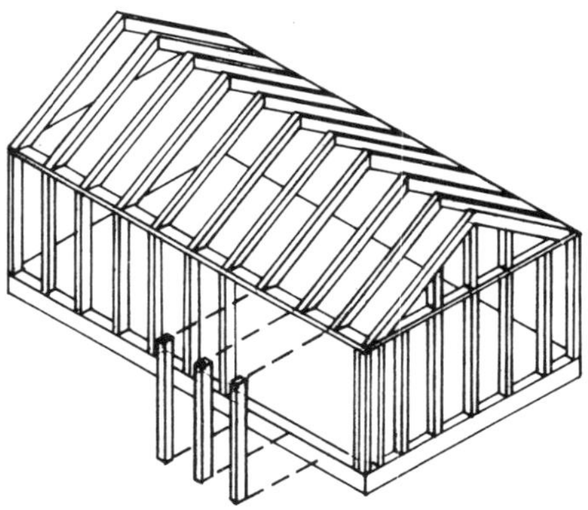

A pre-cut wood building is one in which all the materials needed for construction are factory cut to the proper size, identified, and packaged to be put together at the job site. The modern pre-cut is the descendant of the 'catalogue' houses sold by mail order companies such as Sears Roebuck and Co., which were popular at the turn of the century.

The way in which the pre-cut building differs from conventional wood construction is that in pre-sizing and pre-cutting the lumber, field cutting and material waste are eliminated or reduced. An unskilled laborer in a factory can produce more than a skilled carpenter in the field.

The manufacturers of pre-cut wood buildings usually have stock designs, but if changes or modifications conform to certain rules (usually a 4 foot module), variations are possible. Most pre-cut buildings are for single-family houses, although some other building types are available.

plans

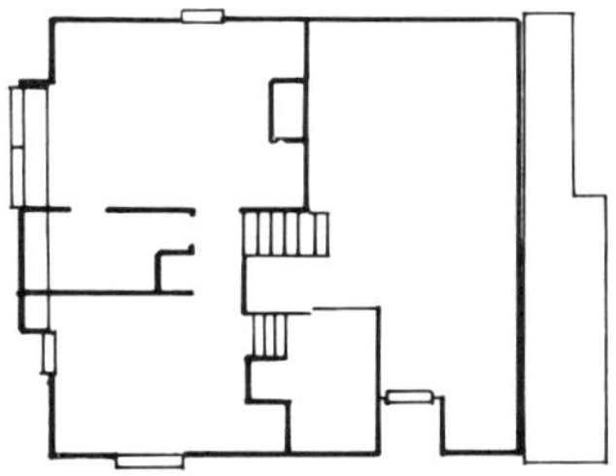

2nd. floor

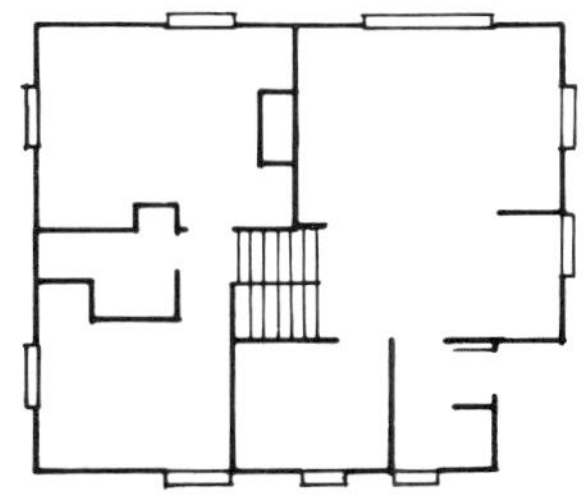

upper level

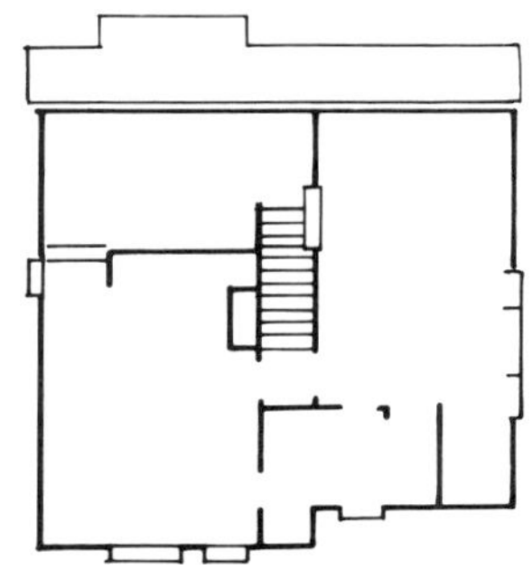

2nd. floor

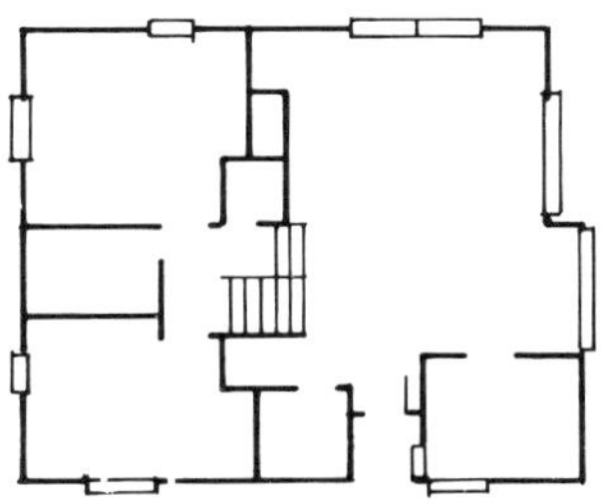

1st. floor

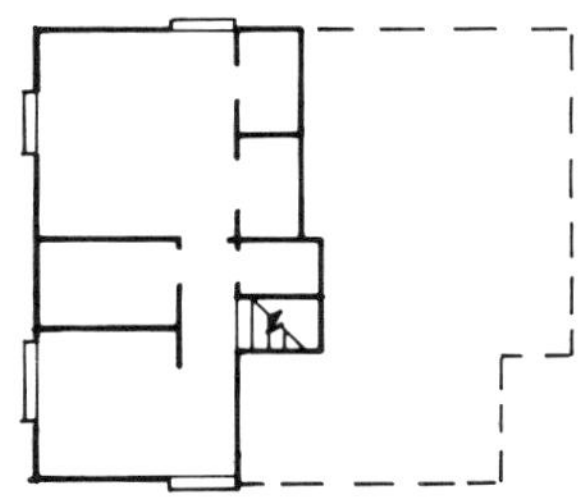

lower level

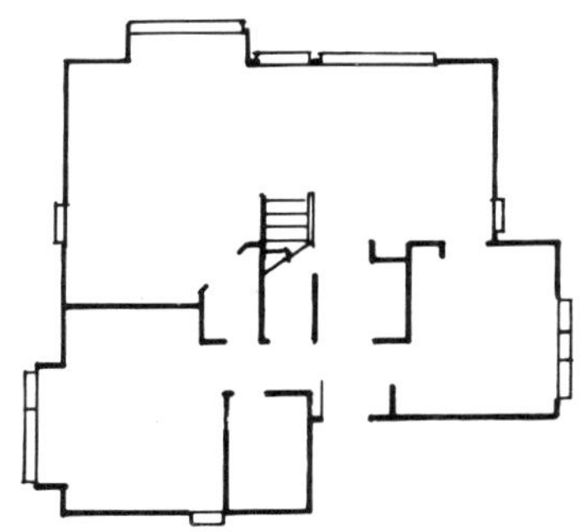

1st. floor

GLAZED

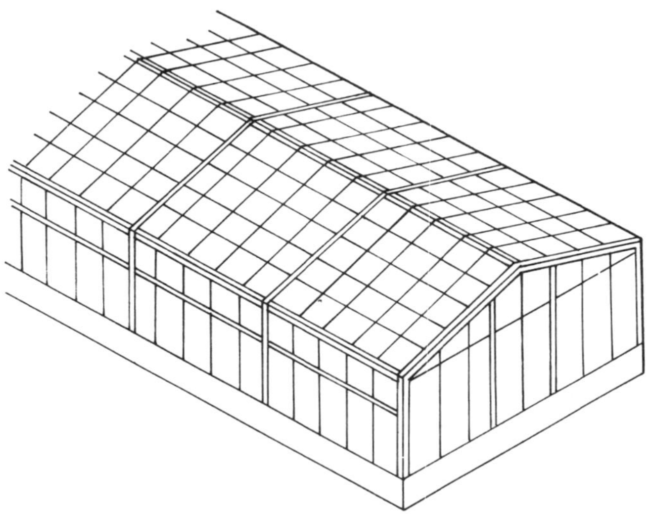

Glazed buildings consist of an assembly of prefabricated metal structure and glass or acrylic enclosure panels — what is generally called a greenhouse. Glazed buildings are one of the oldest industrialized building systems, starting with Paxton's Crystal Palace.

Essentially, a glazed building consists of a steel framework that supports metal purlins and mullions, which in turn support the glazing panels. The steel frame generally consists of mill bents, designed for the spans and loads to be resisted. The purlins and mullions are extruded metal sections incorporating a secondary structure, support of the glazing panels, and moisture protection. The glazing panels of glass or plastic can be single or double layers and can have varying heat and light transmission properties.

The configurations of glazed buildings can be of various shapes and sizes, both in plan and section. They can be designed as separate, free-standing buildings, or in conjunction with and connected to other buildings.

Although the most common example of glazed buildings is the greenhouse, they have also been used for a variety of other applications such as recreation facilities, natatoriums, and industrial facilities.

typical buildings

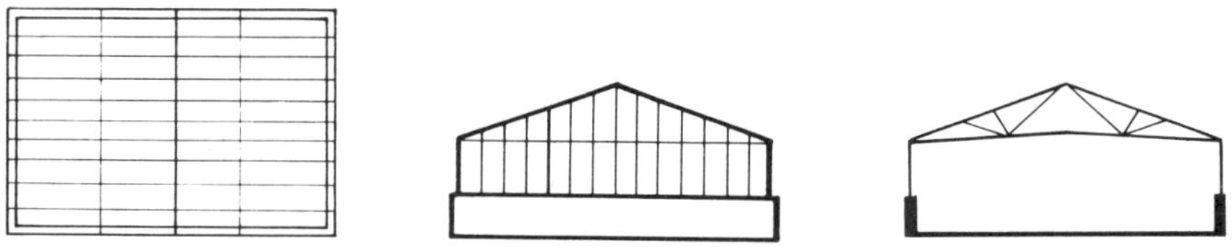

plan elevation section

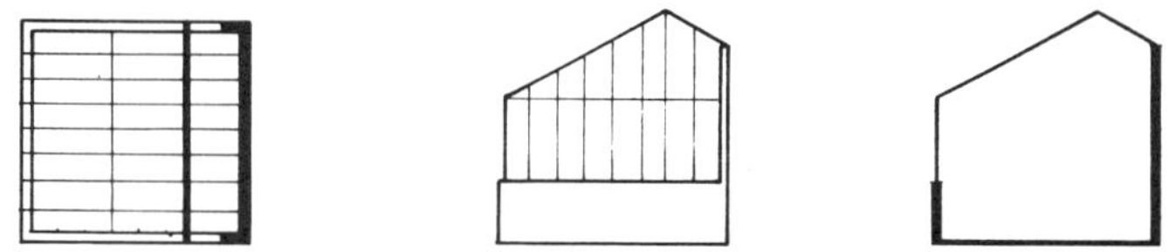

plan elevation section

variations

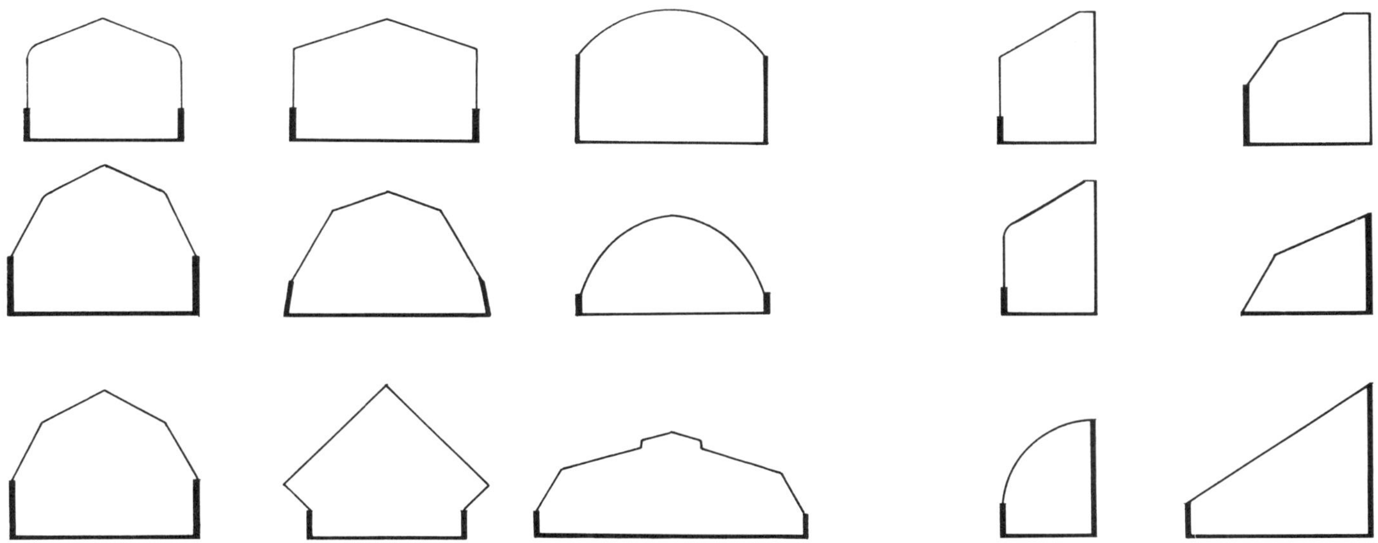

configurations

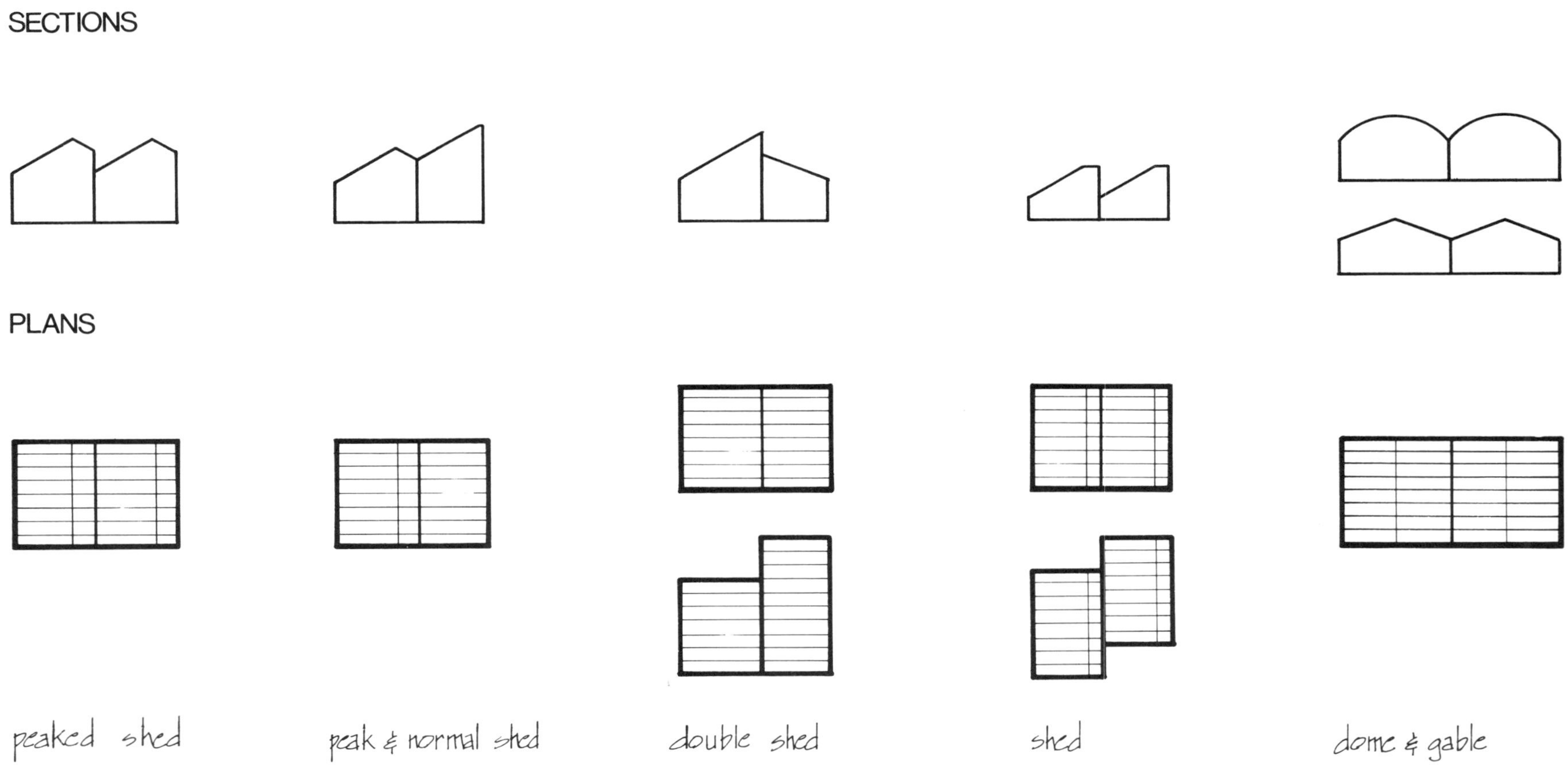

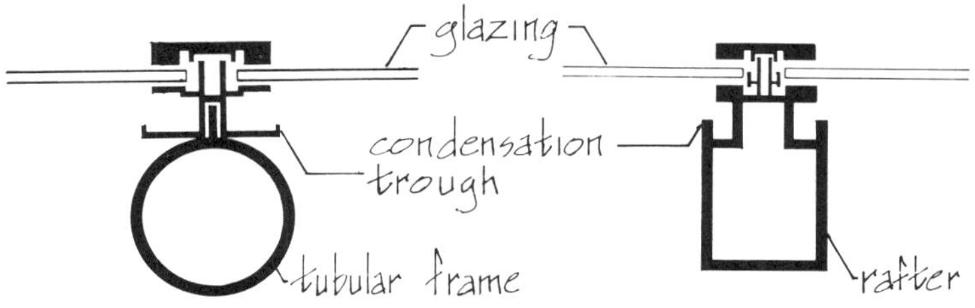

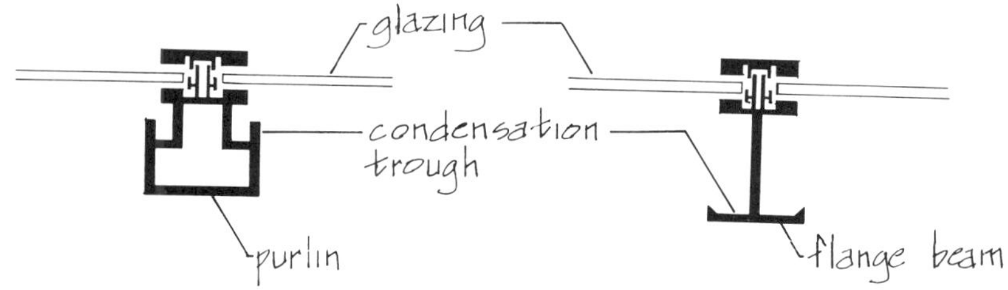

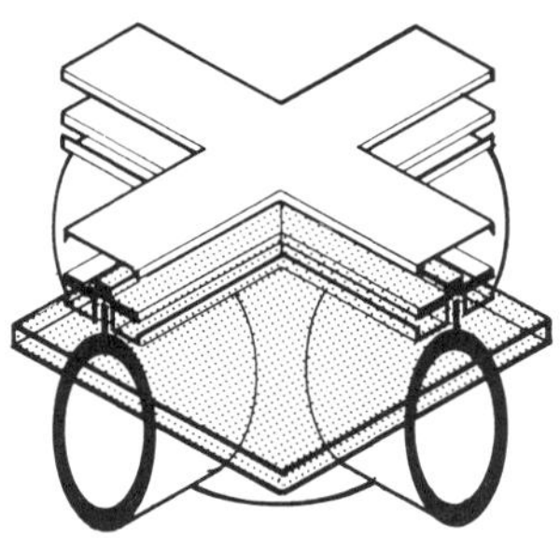

tubular isometric

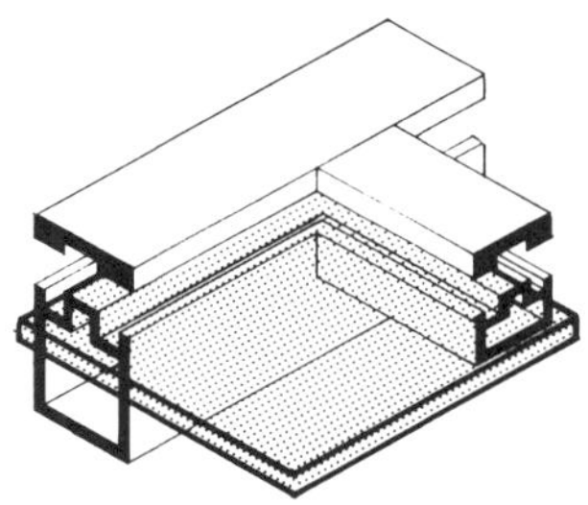

rafter isometric

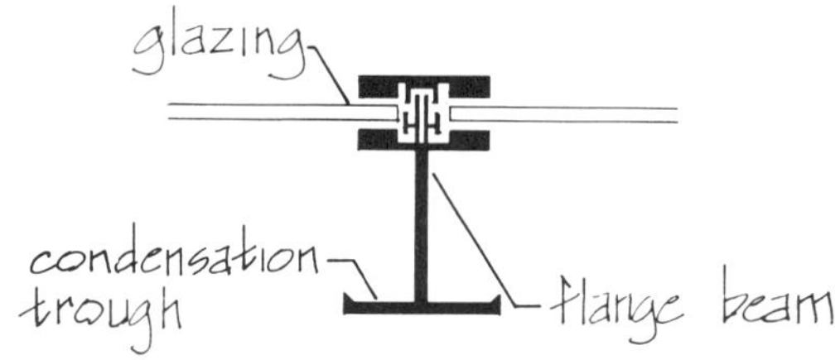

GEODESIC

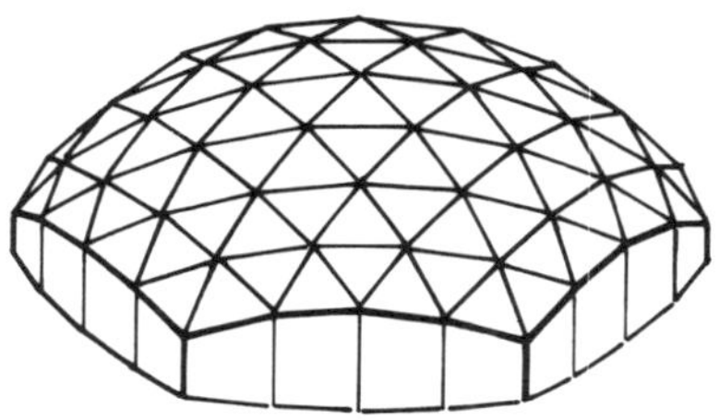

Industrialized geodesic buildings are composed of a steel frame and metal or plastic panels. They provide large, domed, column-free spaces with a minimum of materials. The domed enclosure can vary from a shallow, dish-like surface to a half- or three-quarter sphere.

The steel frame is erected to produce the latticework of triangular subdivisions characteristic of the geodesic. Then, the enclosure panels are fastened to this framework. The enclosure panels can be of metal or plastic and range from transparent to opaque. The opaque panels can also have a variety of surface treatments from an embossed pattern to acoustic panels.

Geodesics have found a wide range of applications including recreational facilities, radar enclosures, auditoriums, and exhibition halls.

configurations

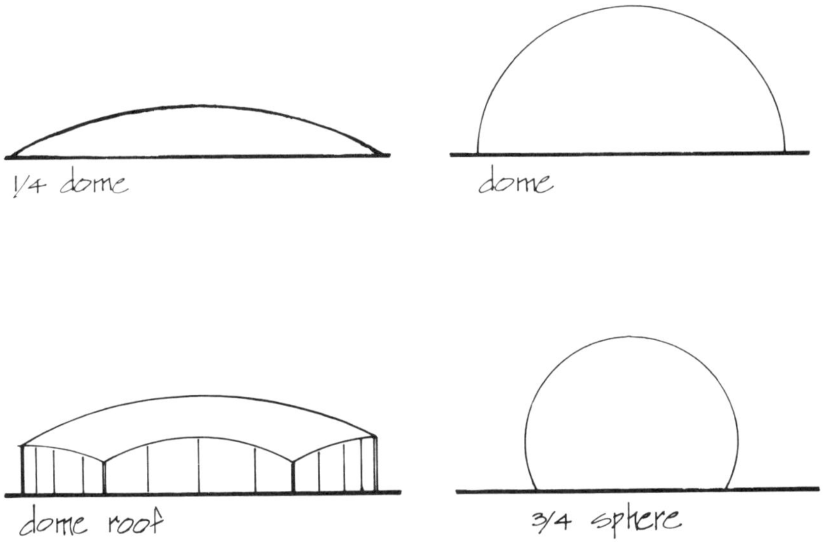

diameter: 50 to 200 feet
height: 7 to 40 feet

details

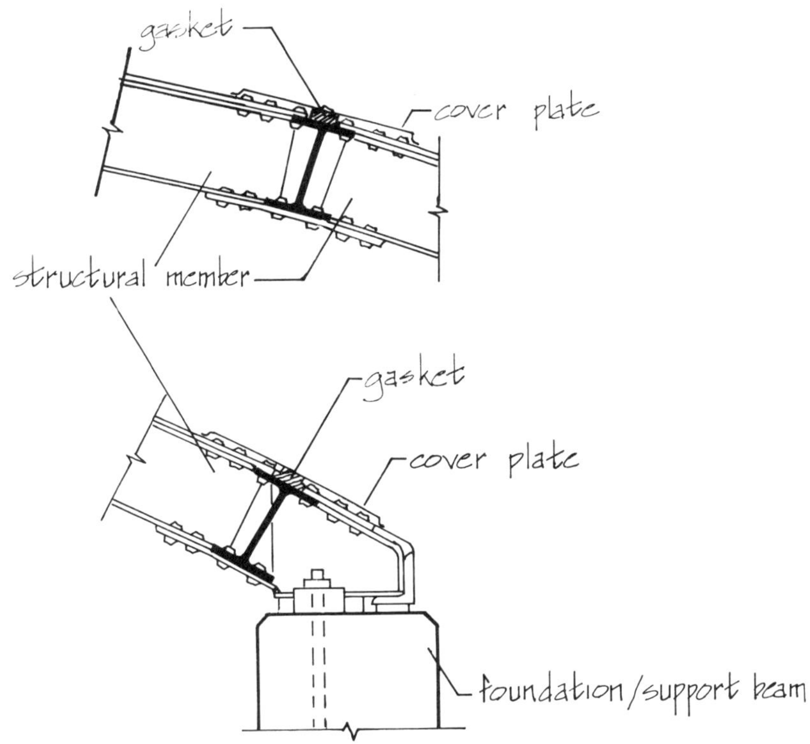

MOBILE HOMES

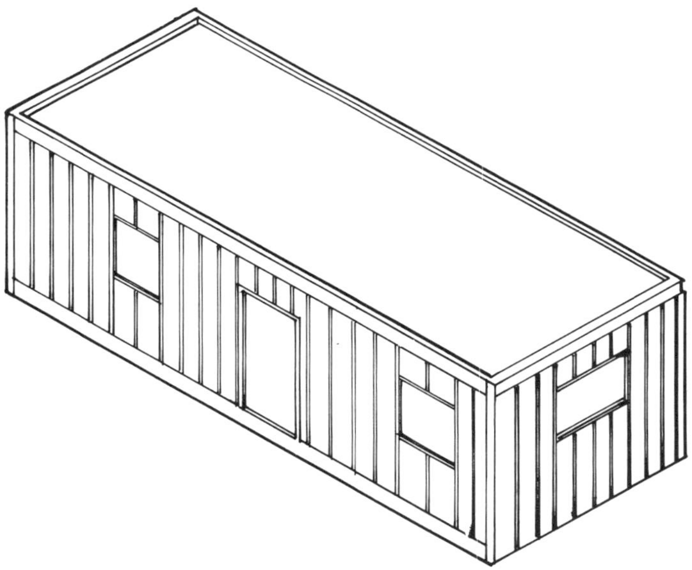

Mobile homes are a truly prefabricated building system. Developed from the trailer homes and campers of the early part of this century, mobile homes are designed to be easily and readily transportable, and to require a minimum of site amenities. The reality that a majority of mobile homes are never moved from their original locations does not change the fact that they have the potential.

Mobile homes generally consist of light wood or aluminum frame mounted to a steel chassis. The exterior is generally clad in aluminum siding and the interior is finished with sheets of plywood paneling. The size of the unit is limited by the restrictions for highway transit imposed by the various states, but by the use of pull-out sections and 'double-wides' (similar to sectionals) larger units can be formed.

Due to its requirement for transportability, the mobile home is constructed of lighter weight materials, but for that same reason, it is not as sturdy as conventional construction.

Mobile homes are much less expensive than conventional construction, due to their efficient design, factory production, and different financing. Many mobile homes offer a complete package including not only the structure but also complete furnishings.

plans

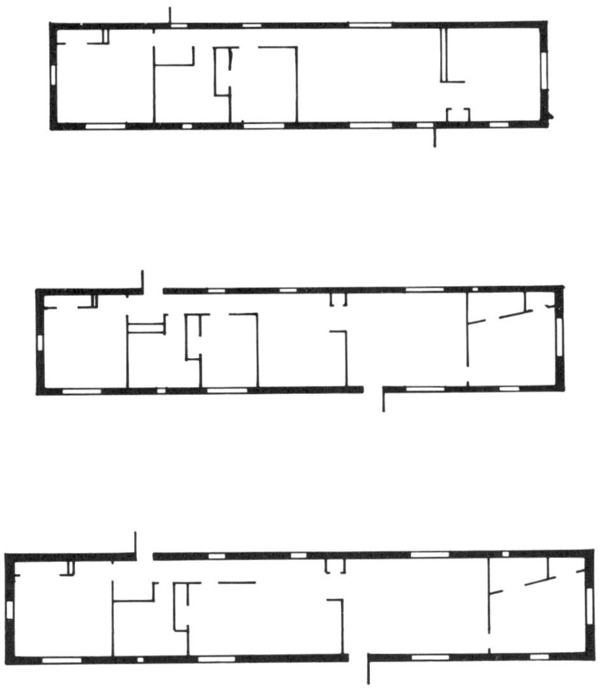

single-width units

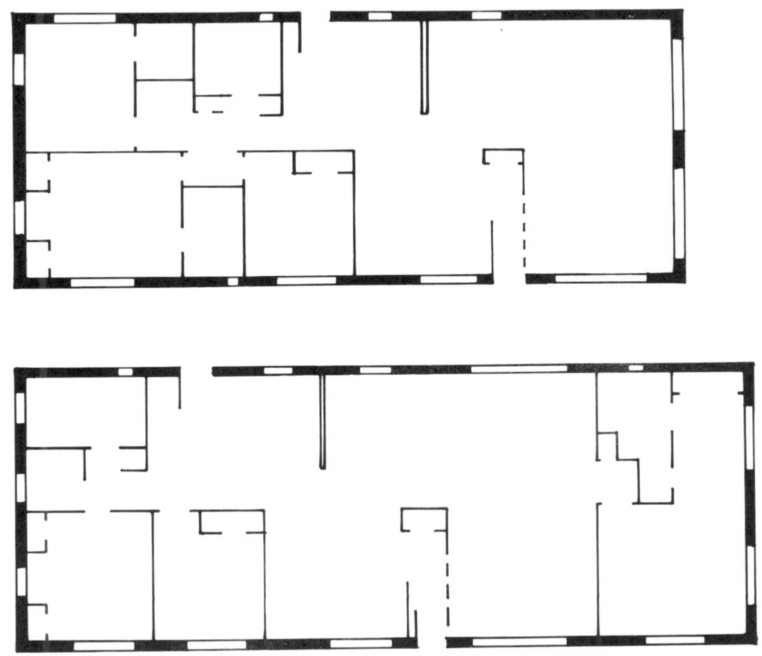

double-wide units

SYSTEMS

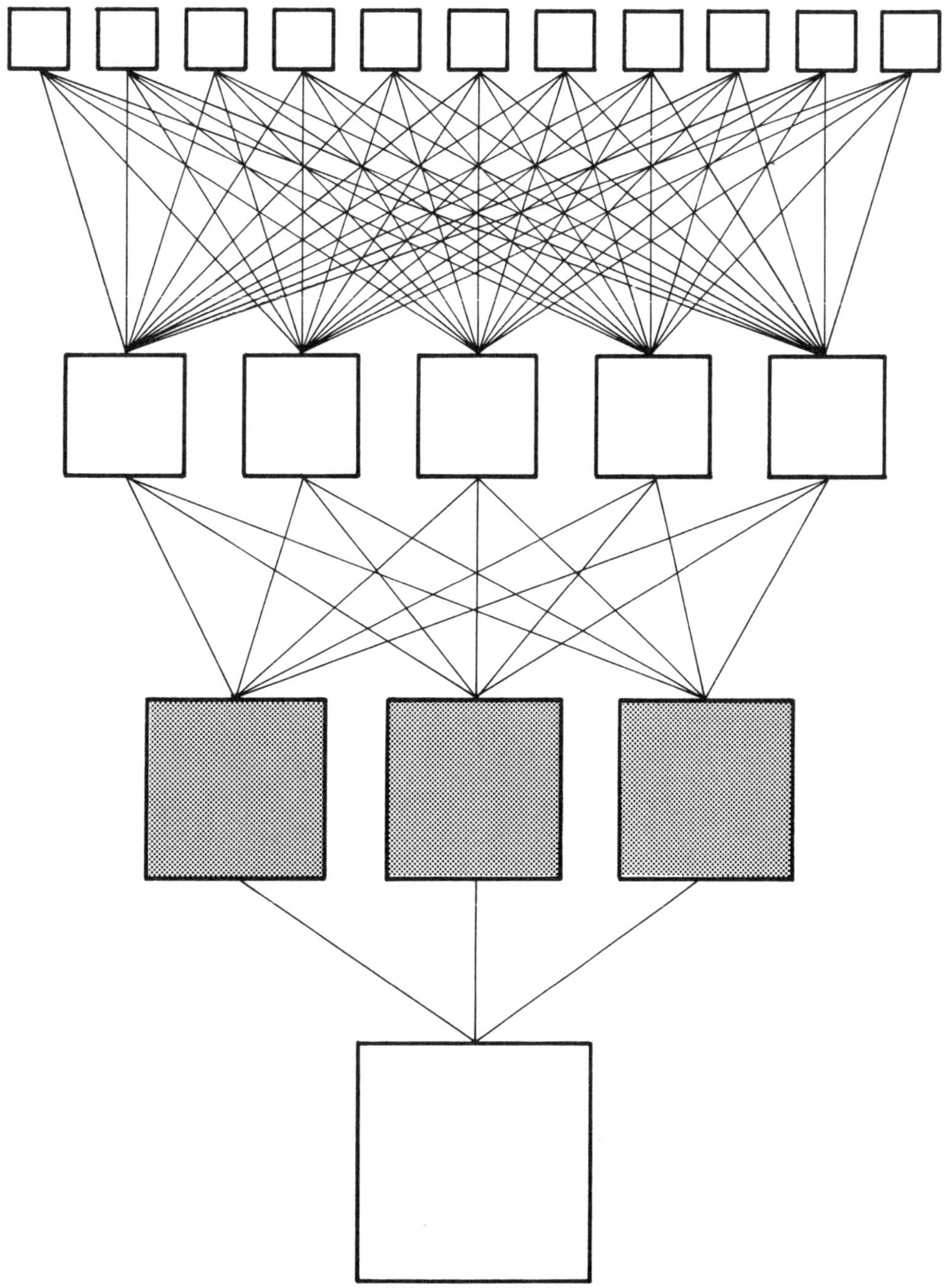

STEEL FRAMING

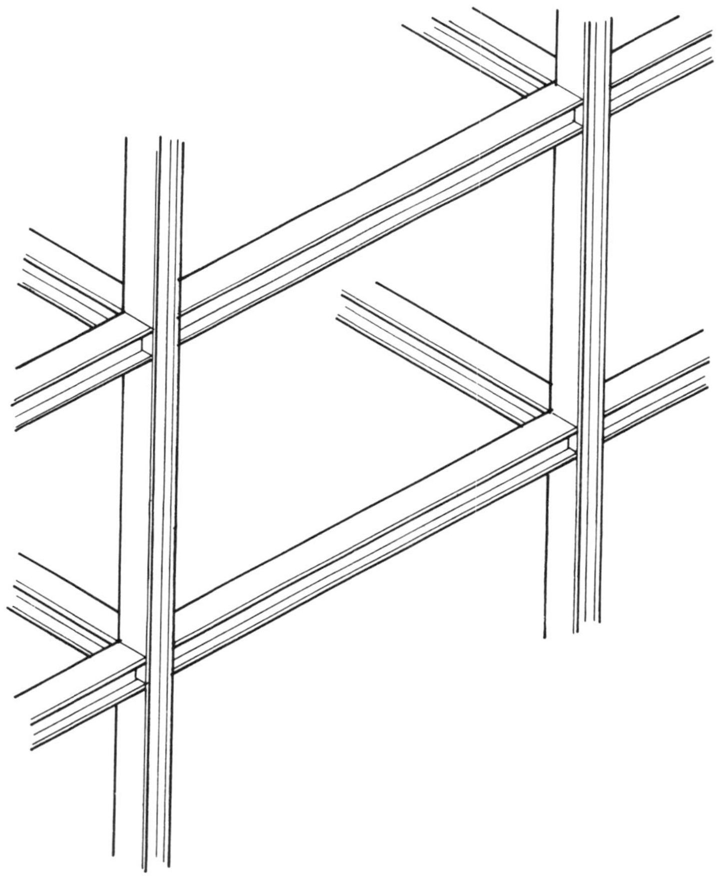

Steel framing is made from structural-grade steel which is factory fabricated into desired shapes and sizes. Shipped to the site, it is erected and fastened together by either bolting or welding to produce a structural skeleton frame on which the enclosure and finish materials are attached. Although in common, everyday usage, steel framing is a prefabricated building element since it is factory produced, and in some cases, even partially preassembled.

A variety of standard cross-sectional shapes are commonly produced and available. Custom designs are also possible to produce any configuration designed. Steel fabricators can also prepare sections of the frame by pre-drilling bolt holes and shop welding gusset plates.

Steel framing can accommodate a wide range of structural requirements from small structures to the largest high-rises.

standard shapes

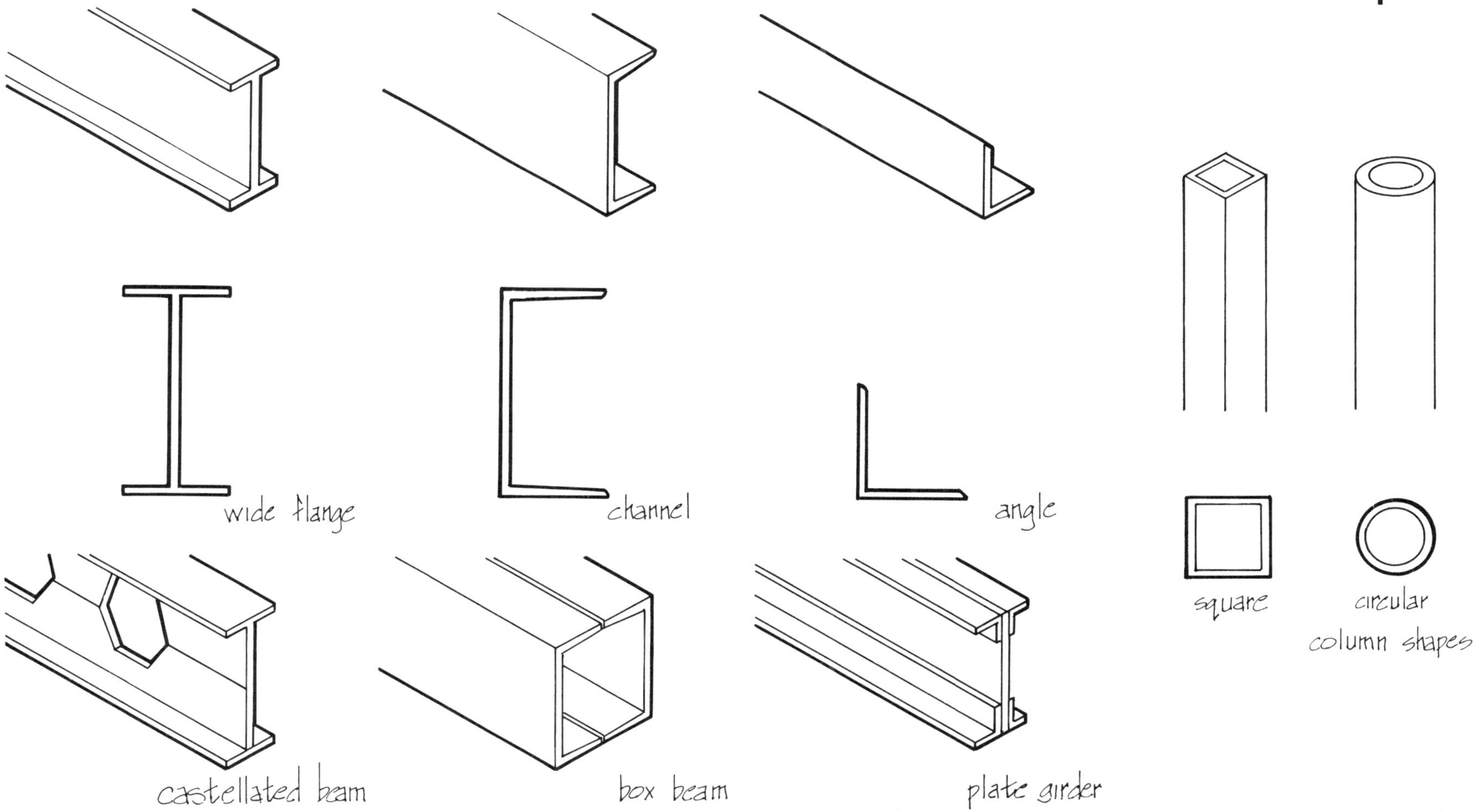

STEEL STUDS

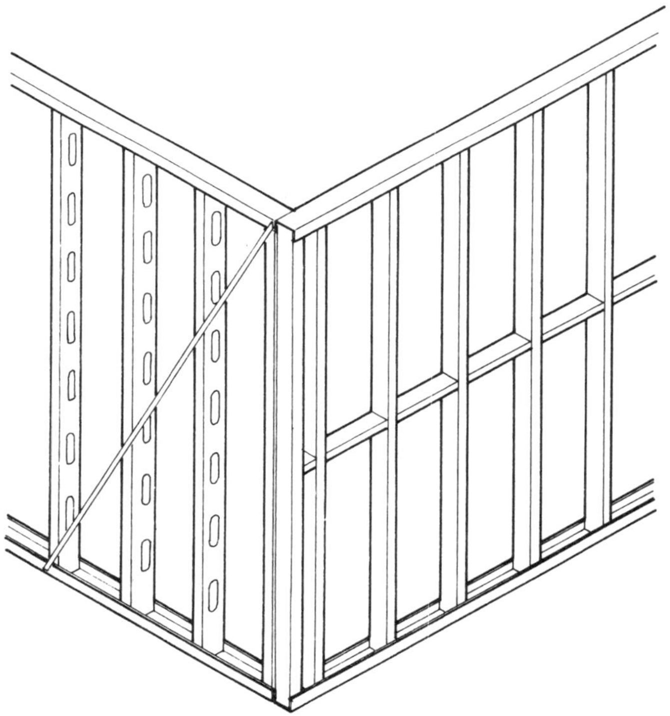

Steel studs represent an attempt to apply metal technology to light framing and non-load-bearing construction. Light-gauge steel is formed into lightweight, easily handled shapes which can be used in place of conventional wood studs. A variety of shapes are available for various applications.

The structural capabilities are slightly greater than wood, allowing centering of studs on 24″ centers. The use and connection details are extremely similar to those used for wood construction. Steel studs have the additional advantages of dimensional stability, resistance to biological decomposition, and uniform properties. Comparative cost, labor, and building weight figures are more favorable with steel than with wood.

Steel studs can be used for structural purposes, usually in conjunction with steel joists, to provide the primary structure, such as in house construction. They can also be used as a secondary enclosure or partition wall system, such as in high-rise office buildings.

Various manufacturers provide a good selection of types and sizes of steel studs, generally sufficient to meet the needs of a variety of project applications. The studs can be solid or may have holes provided which allow conduit and other services to be run through them without field cutting and reducing of the structural strength. Finish materials can be fastened in several ways, most commonly by self-drilling screws, or spiral shank nails.

types & sizes

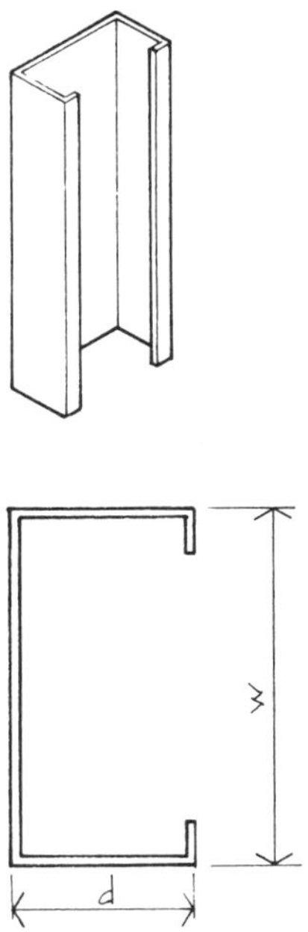

cee stud

	typical sizes
w	2½", 3⅝", 4", 6"
d	1⅝"
length	6'-0" min.; 40'-0" max.

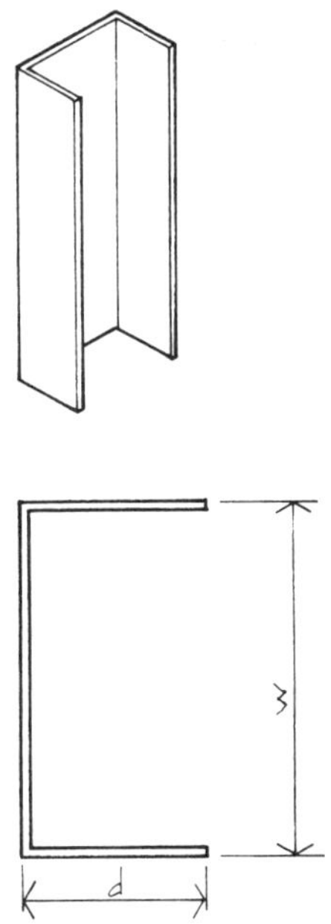

channel stud

	typical sizes
w	2½", 3⅝", 4", 6"
d	1", 1⅜"
length	6'-0" min.; 40'-0" max.

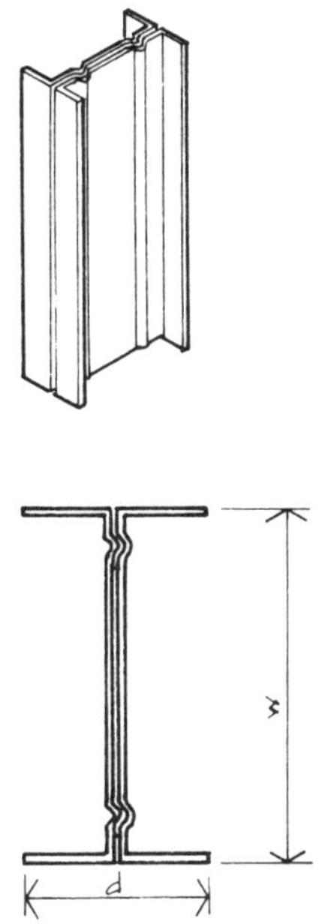

nailable stud

	typical sizes
w	2½", 3⅝", 4"
d	1¾", 1 13/16", 1 15/16", 2¼"
length	6'-0" min.; 40'-0" max.

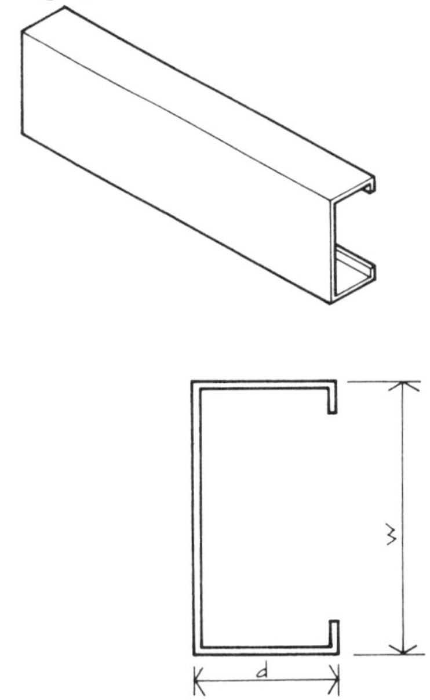

lintel

	typical sizes
w	2 11/16" to 12 3/16"
d	¾", 1½", 1¾"
length	20'-0"

section details

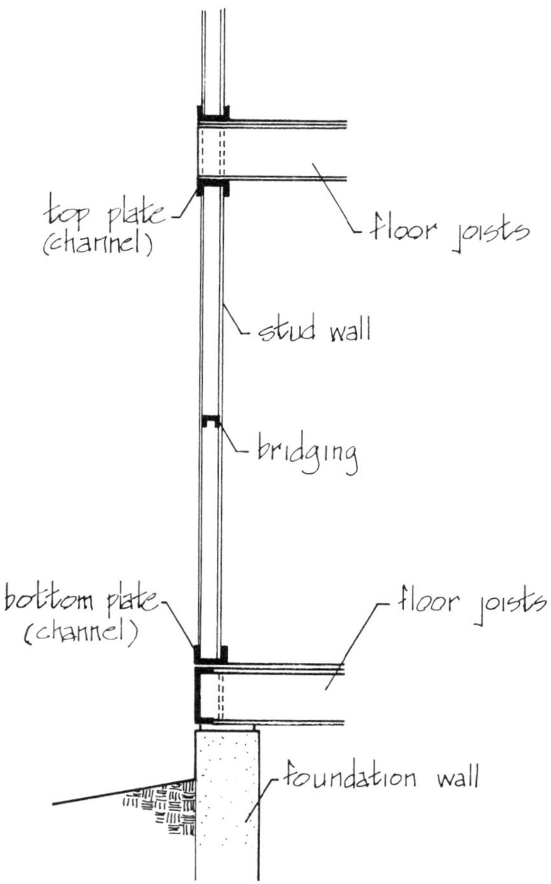

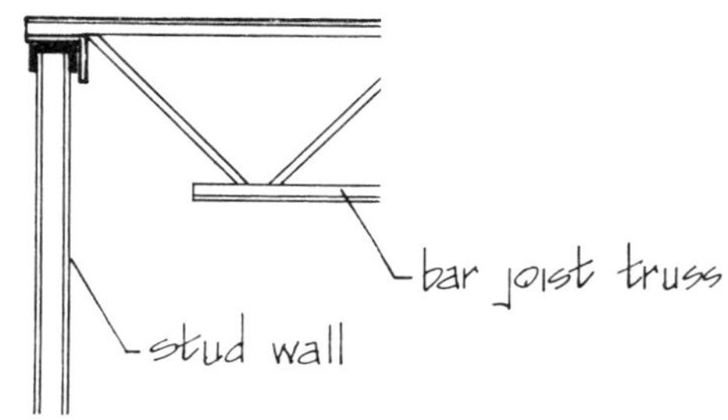

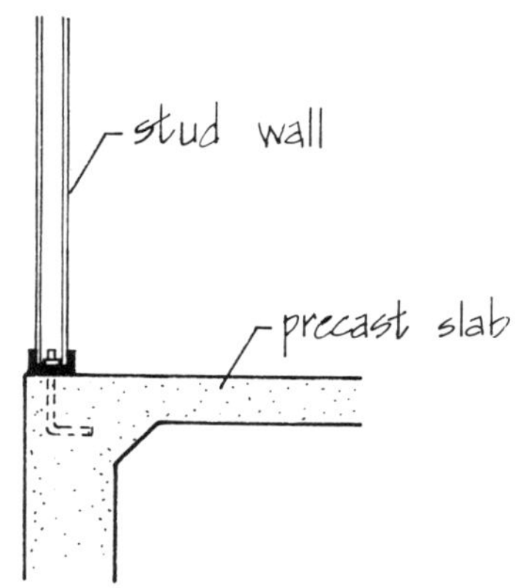

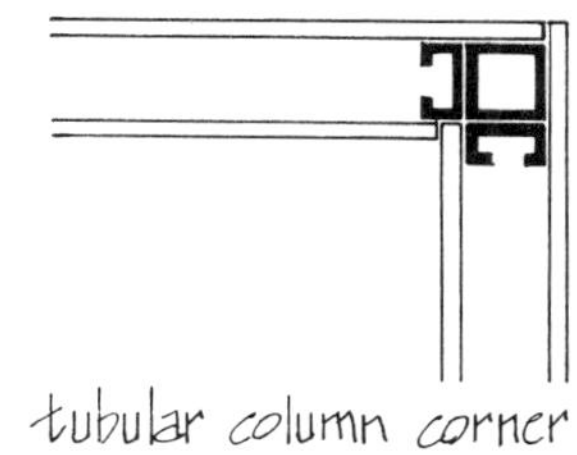

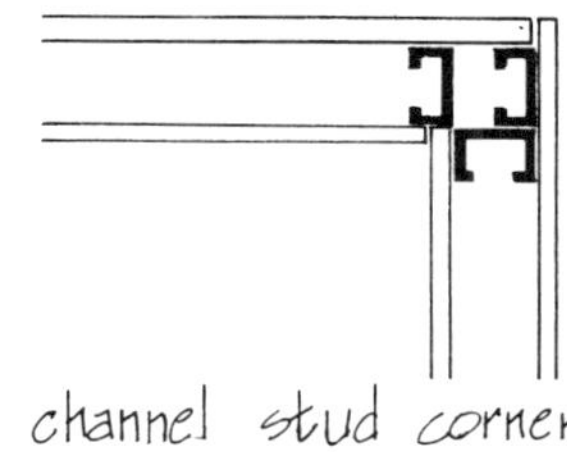

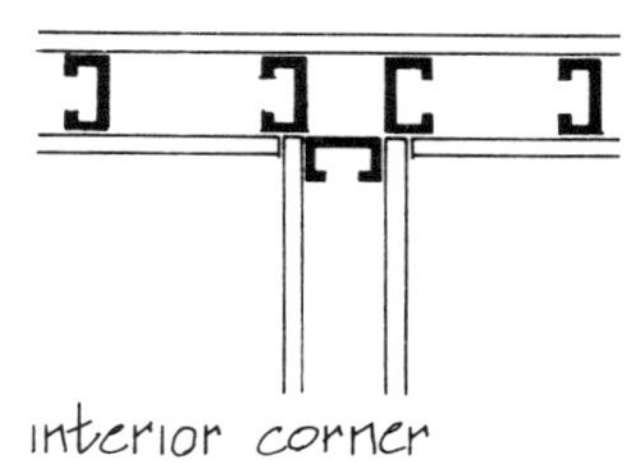

elevation

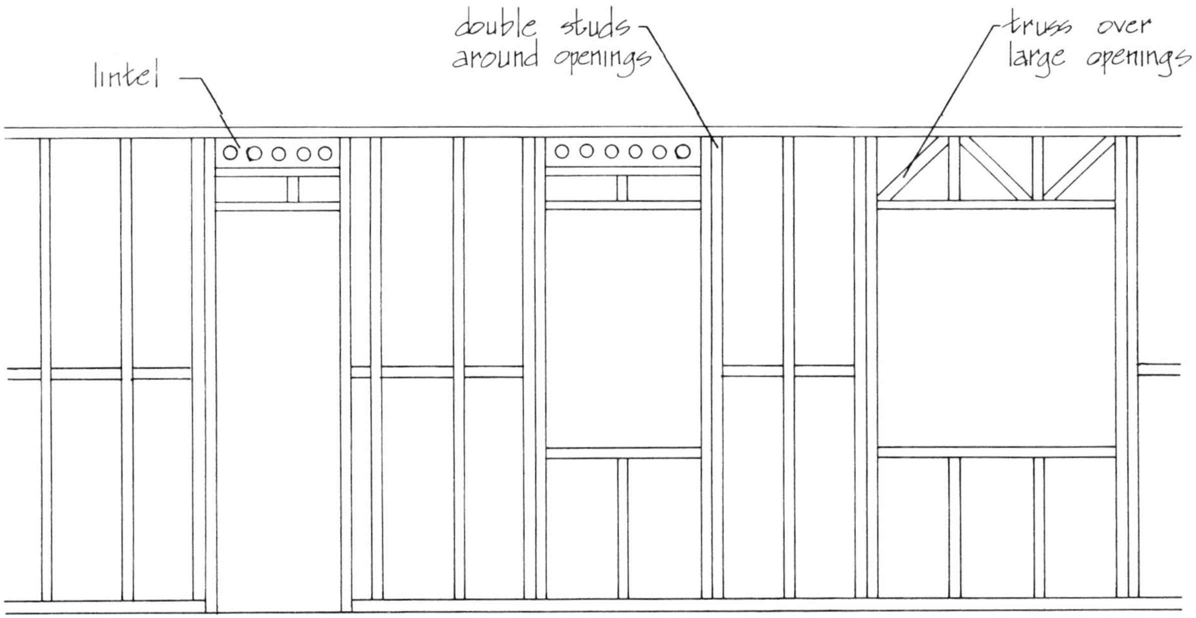
lintel
double studs
around openings
truss over
large openings

STEEL JOISTS

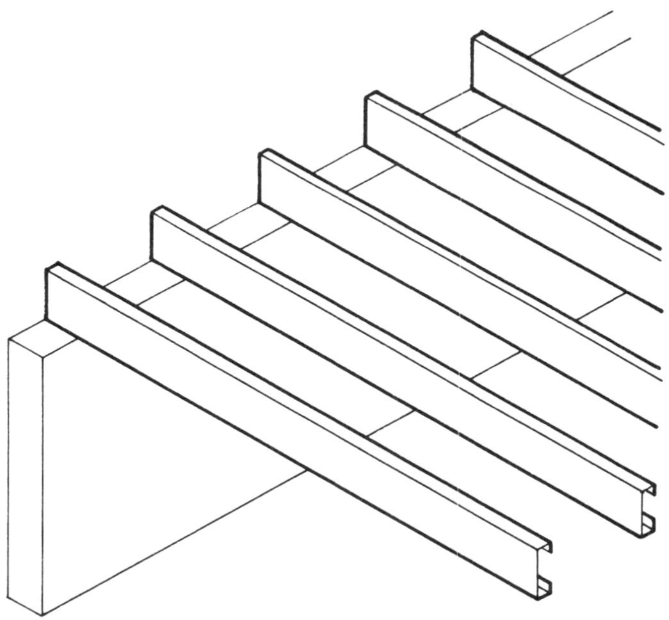

Steel joists are made from galvanized steel in a variety of sizes and shapes to be used in place of conventional wood joists. When used in conjunction with steel studs they can provide a total light framing system. They may also be used in conjunction with other materials such as masonry block.

Steel joists are manufactured in standard depths and lengths with appropriate structural data to allow accurate structural design. They are available with solid web or with holes punched in the web which allow mechanical services to pass through them without reducing their structural strength. Their application is very similar to that of wood joists, and fastening is generally performed by the use of metal screws. Finish materials are attached by means of self-tapping screws or spiral shank nails.

Steel joists have found applications in a variety of one- to three-storey residential and commercial construction. They compare favorably with wood construction, and have the added advantages of being dimensionally stable, fire and rot resistant, and possessing uniform properties.

types & sizes

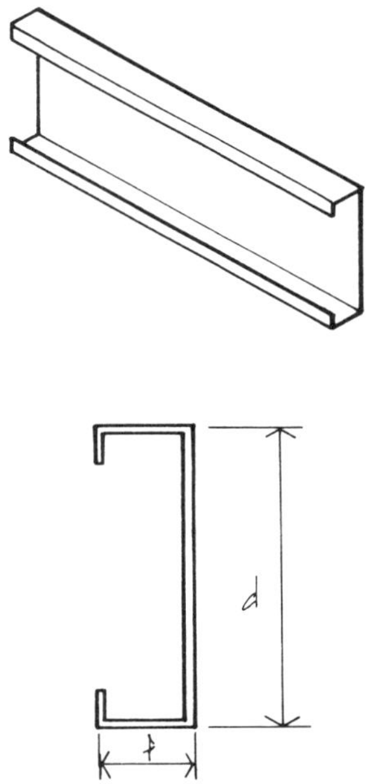

cee joist

	std. sizes
d	5½", 6", 7¼", 8", 9¼", 10", 12"
f	1¾", 1⅞", 2½"
gauge	12, 14, 16, 18, 20

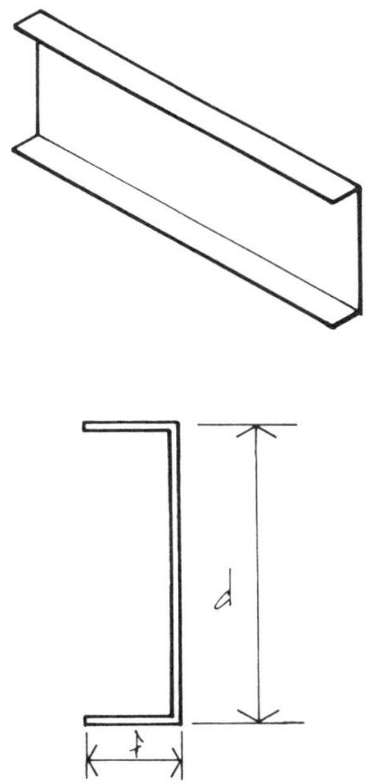

channel joist

	std. sizes
d	6 3/16", 8 3/16", 10 3/16", 12 3/16"
f	¾", 1⅞"
gauge	14, 16

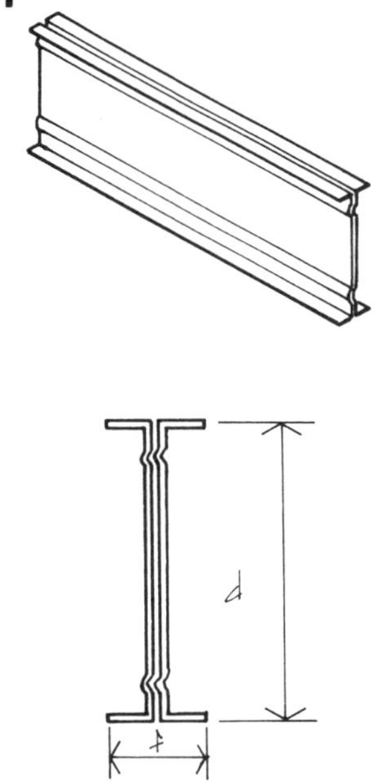

nailable joist

	std. sizes
d	6", 8", 9", 10", 12"
f	1⅞", 1 15/16", 2¼", 2⅜", 2½", 2⅞"
gauge	12, 14, 15, 16, 18

hardware

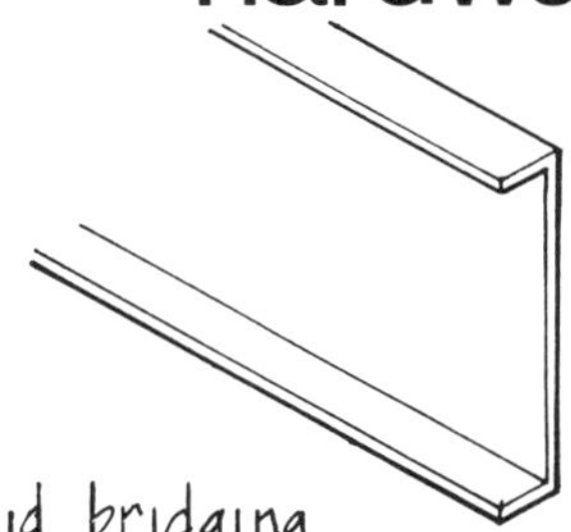

solid bridging

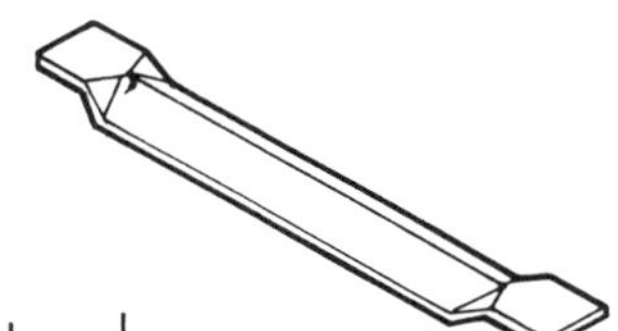

v-bridging

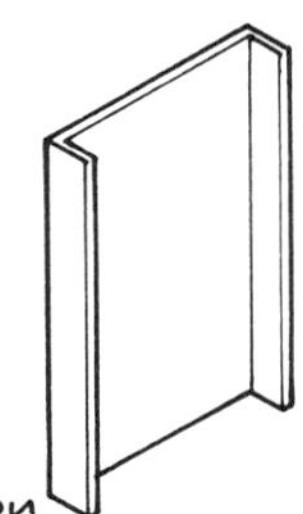

web stiffener

assembly

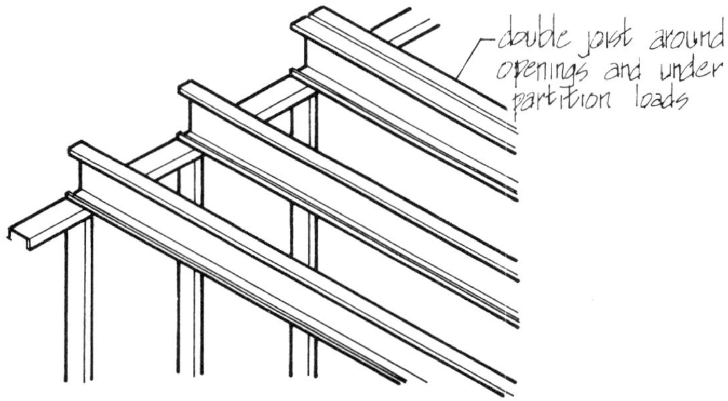

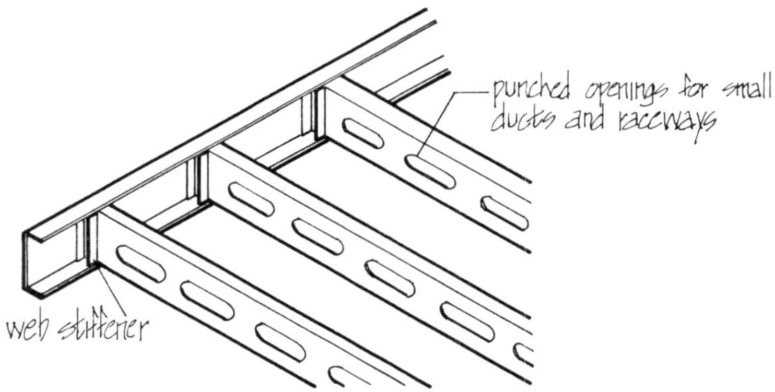

details

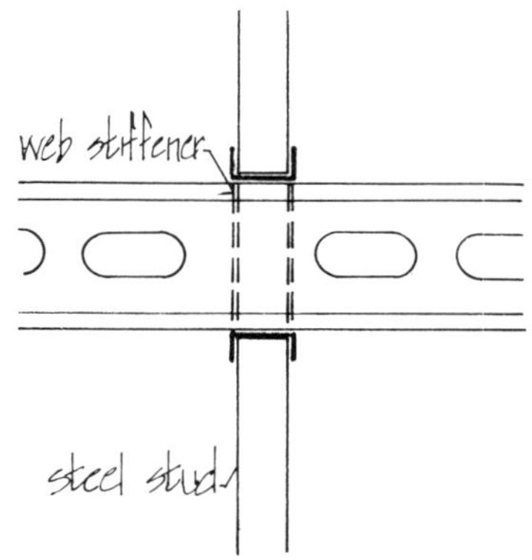

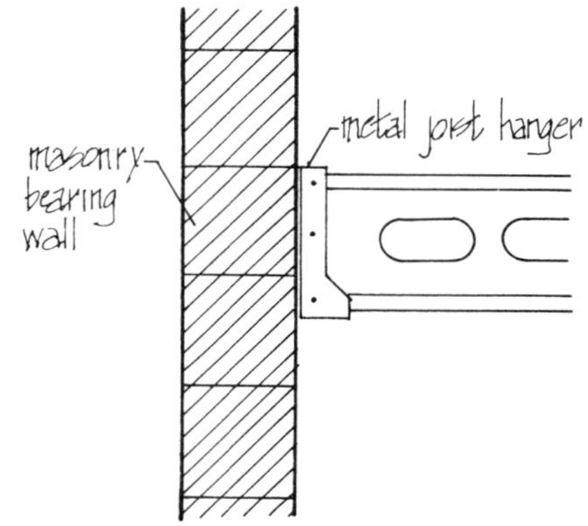

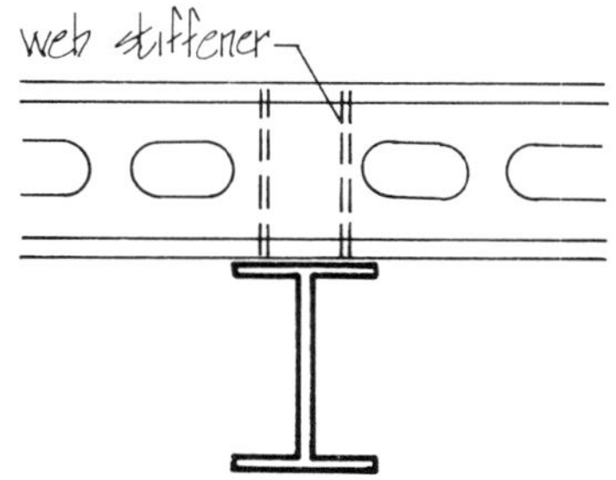

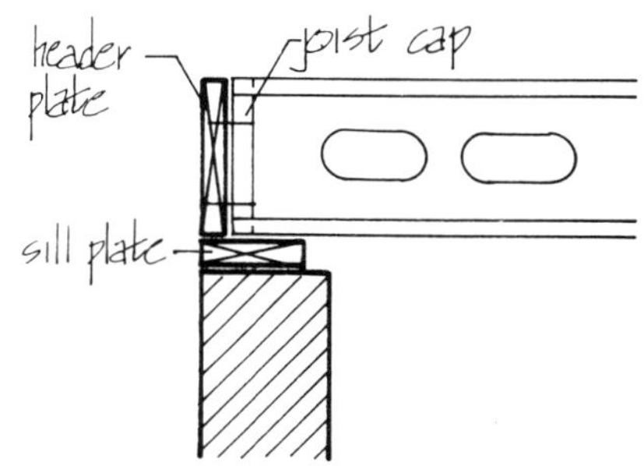

span table

allowable uniform load

joist size (depth)	type span	spacing	span in feet 10	12	14	16	18	20	22
6"	S	16"	114	52	33	28			
		24"	76	35	22	19			
		48"	38	17	11	9			
	C	24"	143	54	41	34			
		48"	71	27	20	17			
8"	S	16"	233	135	85	57	53		
		24"	156	90	57	38	34		
		48"	78	45	28	19	17		
	C	24"	140	117	100	72	50		
		48"	70	58	50	36	25		
9"	S	16"			114	76	54	39	
		24"			76	51	36	26	
		48"			38	26	18	13	
	C	24"			92	81	67	49	
		48"			46	40	33	24	
10"	S	16"			186	125	87	64	48
		24"			124	83	58	43	32
		48"			62	42	29	21	16
	C	24"			134	117	105	80	60
		48"			67	59	52	40	30
12"	S	12"						163	23
		24"						109	82
		48"						54	41
	C	24"						186	154
		48"						93	77

PRECAST COLUMN & BEAM

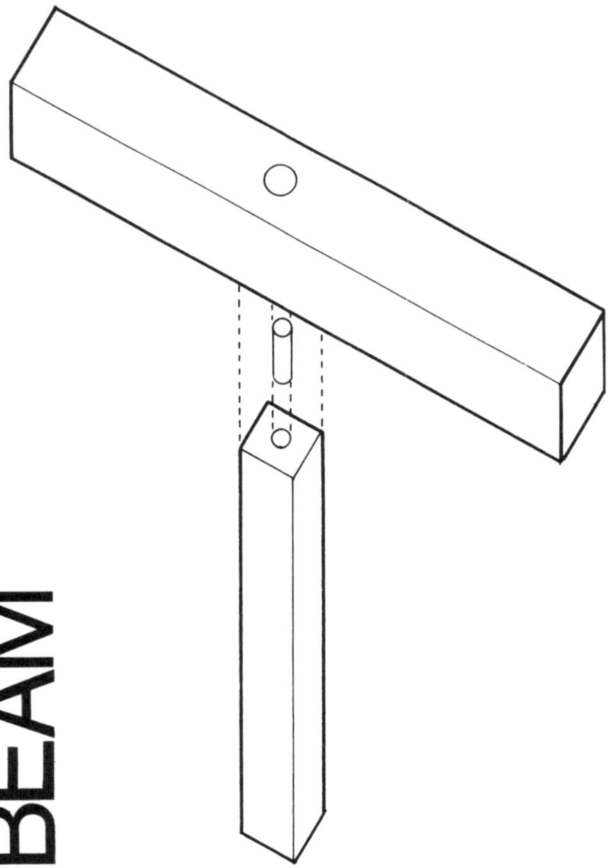

Precast concrete column and beam systems provide a rigid structural framework in less construction time than poured-in-place concrete. They consist of a coordinated system of precast columns and precast beams which are simply erected and fastened together at the job site.

Standardized column and beam sizes and shapes are factory produced and can be combined to produce the primary structural system. Secondary structure and enclosure systems can then be applied to this framework.

The system basically consists of column elements which have a dowel or threaded insert set in the end which connects to the beams and column above. After the reinforcing steel of the parts are connected together, the joint is grouted to provide a mechanical bond.

The precast concrete column and beam system allows for faster construction time, stronger and more uniform concrete properties, and minimal field labor.

sizes

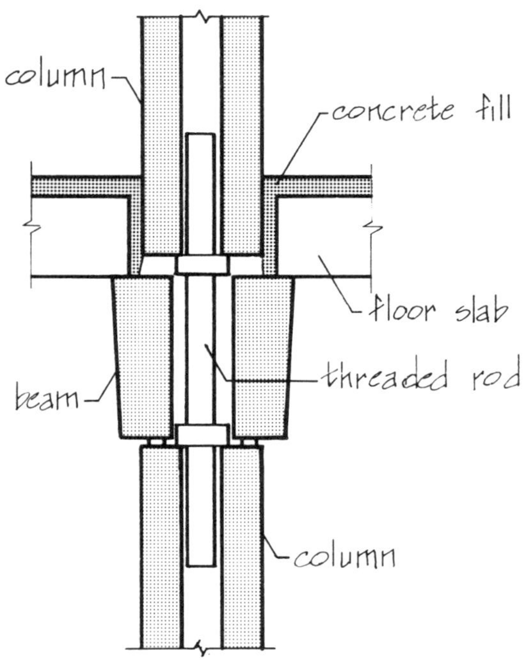

beam sizes

width/depth			
12" × 8"	16" × 8"	20" × 8"	24" × 8"
12" × 12"	16" × 12"	20" × 12"	24" × 12"
	16" × 16"	20" × 16"	24" × 16"
		20" × 20"	24" × 20"
			24" × 24"

column sizes

width/depth			
12" × 12"	16" × 16"	20" × 20"	24" × 24"

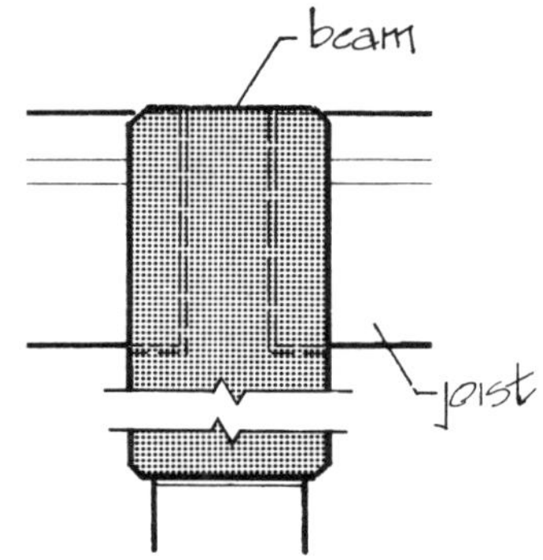

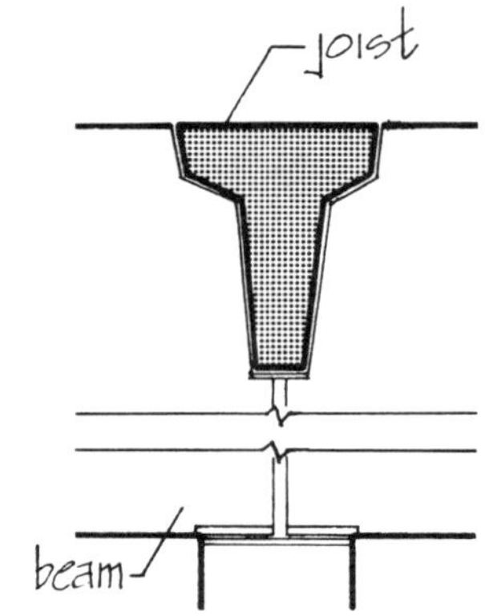

STRESS-SKIN PANELS

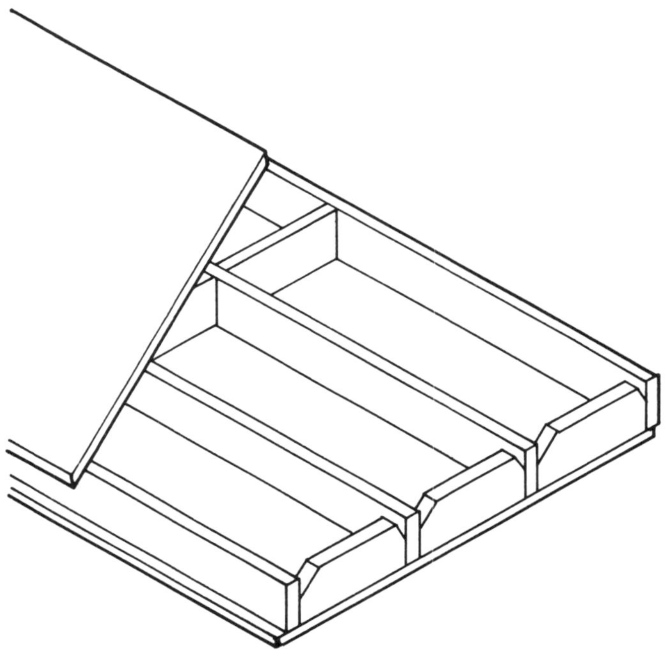

Stress-skin panels are composed of plywood panels which are glued to longitudinal framing members. This construction produces panels which are simply shipped to the job site and installed.

By the use of the glue joint, the entire panel element acts as a single structure in resisting loads. The membrane action of the plywood is utilized in determining the strength of the panel. Structurally, the action of a stress-skin panel is similar to a series of adjoining built-up wooden I beams.

The stress-skin panel is economical not only due to the reduced field erection time, but also because the plywood faces can serve as exterior siding and sheathing and finished interior surface.

They can be used for floors, walls, or roofs, and have applications in all types of construction, including residential, institutional, commercial, and industrial.

sizes

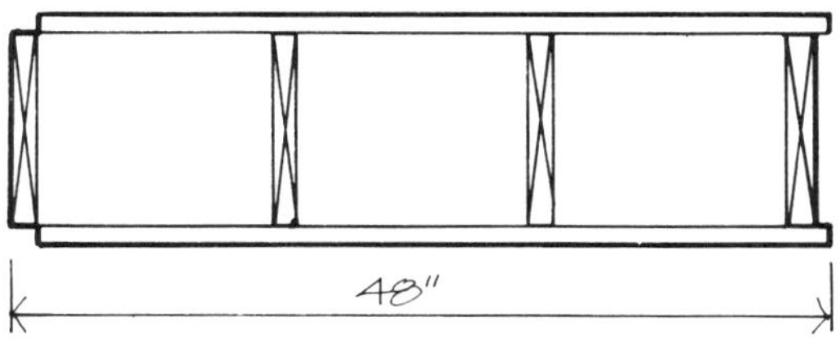

allowable spans

panel depth	total load (psf)						
	20	25	30	35	40	45	50
4¼"	18'-11"	16'-11"	15'-5"	15'-4"	14'-4"	13'-6"	12'-10"
6⅛"	25'-1"	22'-6"	20'-6"	20'-4"	19'-0"	17'-11"	17'-0"
8⅛"	31'-5"	28'-1"	25'-8"	25'-5"	23'-10"	22'-6"	21'-4"

configurations

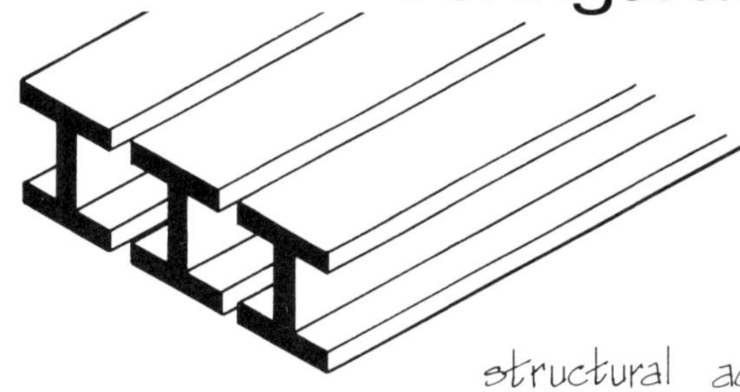

structural action

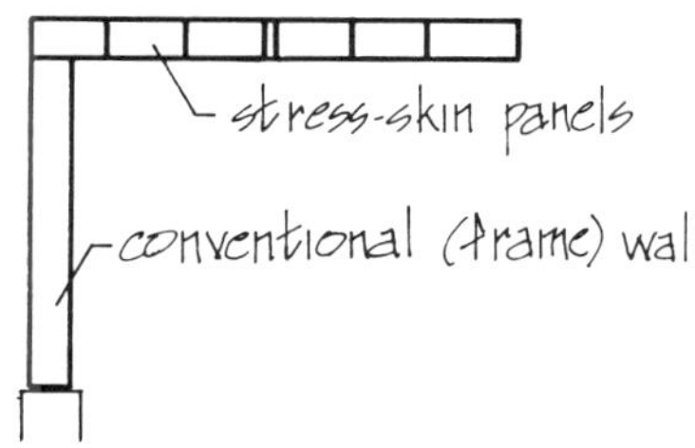

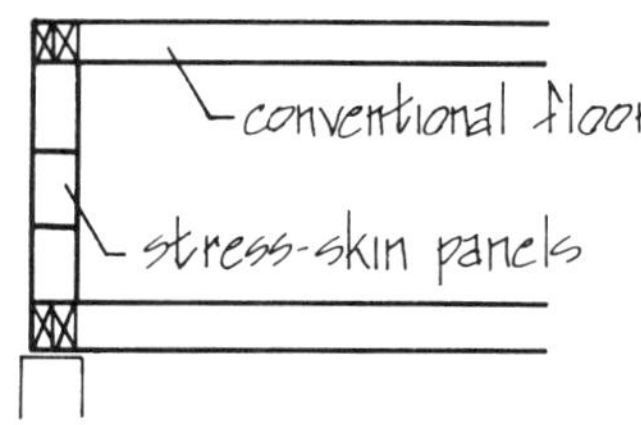

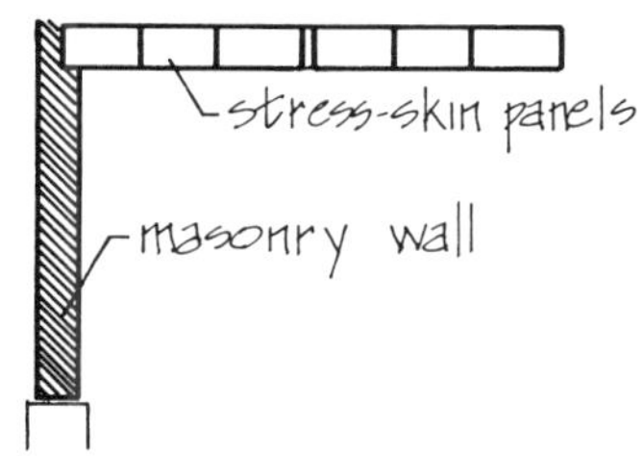

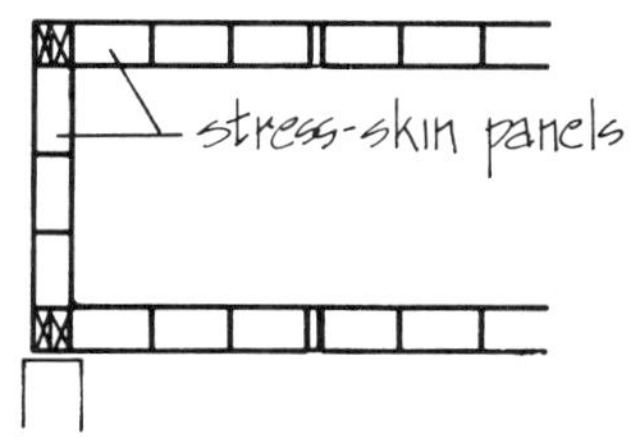

METAL PANELS

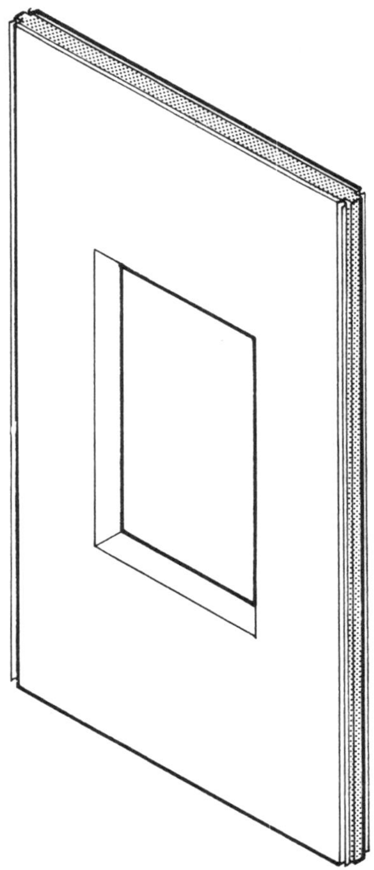

Metal sandwich panels are nonstructural enclosure components. They can be used as the primary enclosure element on steel or concrete framed structures in new construction or as a refacing panel in the renovation of existing masonry buildings.

The panels generally consist of two sheets of aluminum with an inner core of insulating material. The inner skin can also be of some other material, such as plywood or cement-asbestos board. A wide variety of panel configurations can be manufactured, forming virtually any three-dimensional shape desired. Rounded or square corners of any degree or radius, soffits and fascia incorporated in the panel, and square or rounded cutouts are some of the options available. Panel sizes may range up to 5 feet in width and 18 feet in length.

The exterior panel face may be finished in several different ways: natural aluminum, baked-on painted colors, electrolytically coated oxides, and embossed patterns. The openings may be glazed by either standard metal frames or by inset gaskets and glazing. There is also a wide range of accessories available for specific installation conditions (gaskets, reinforcing inserts, copings, and sheet-formed corners), thus facilitating the interface of these elements with other building systems.

Panels are relatively lightweight and easily handled, leading to quick installation at the job site. The panels are available nationwide through a variety of manufacturers and authorized suppliers. Several companies also provide a total systems approach in the marketing of their product that involves design, engineering and detailing, fabrication, and installation. Metal sandwich panels have found applications in many low- and moderate-rise structures, such as hospitals, schools, factories, and office buildings.

panel types

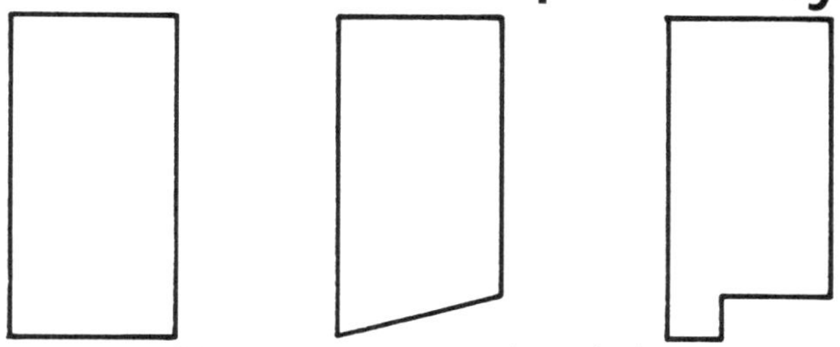

flat — skew cutout — corner cutout

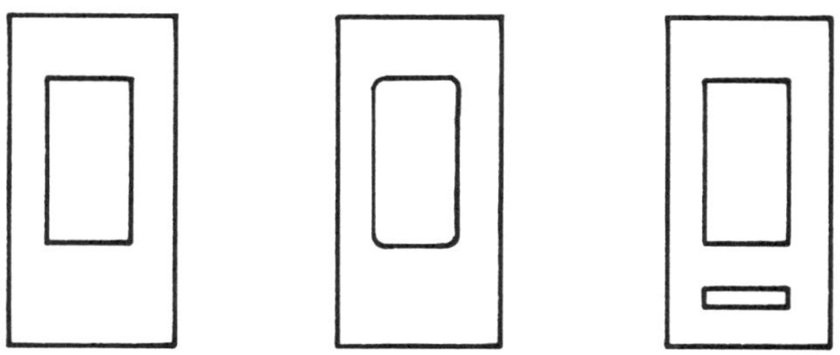

square window cutout — rounded window cutout — window & louver cutout

LOAD/SPAN TABLE

alum. thickness	panel thickness					max. comp. load
	2"	3"	4"	5"	6"	
.025"	9'-9"	15'-6"	18'-0"	18'-0"	18'-0"	400 psf
.032"	9'-2"	15'-0"	18'-0"	18'-0"	18'-0"	1200 psf
.040"	8'-4"	4'-4"	18'-0"	18'-0"	18'-0"	1450 psf

edge conditions

butt joint — single recess — double recess

U-FACTORS

panel thickness	core material	
	polystyrene	polyurethane & isocyanurate
1"	.206	.133
2"	.113	.071
3"	.078	.048
4"	.059	.036
5"	.048	.029
6"	.040	.025
7"	.035	.021
8"	.030	.018

typical joints

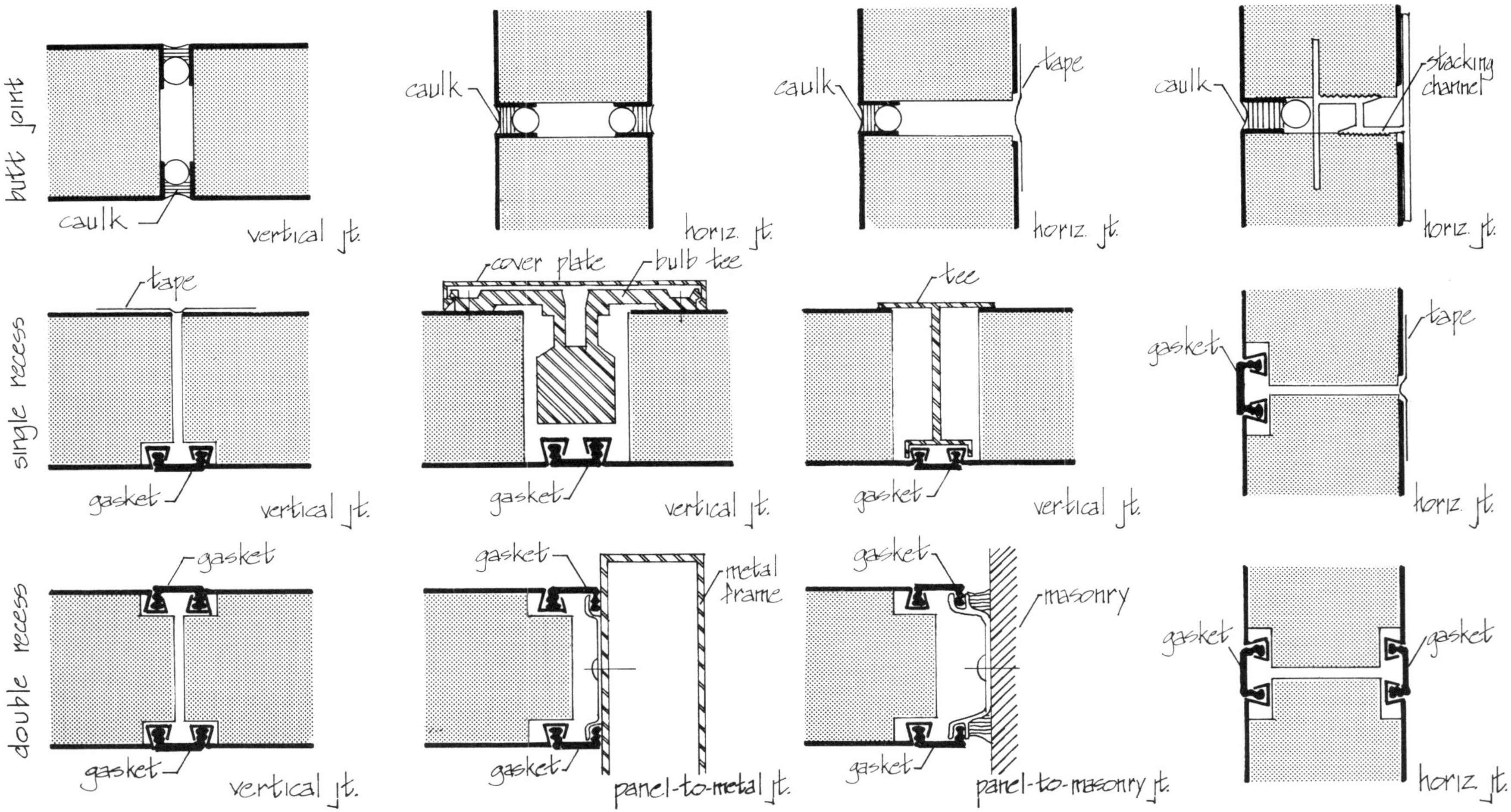

CORNER / SILL

CONNECTIONS

WINDOWS

details

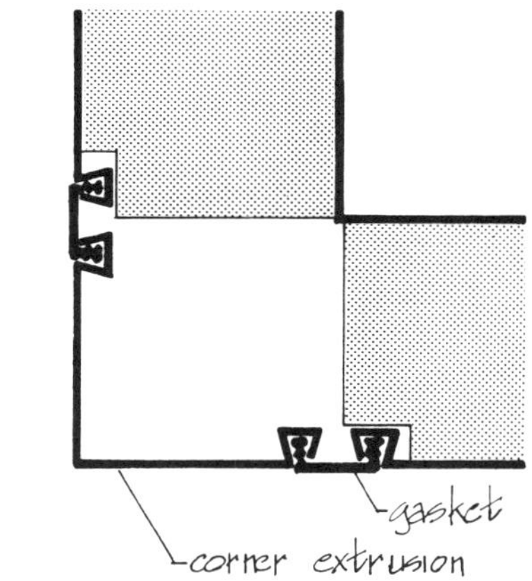

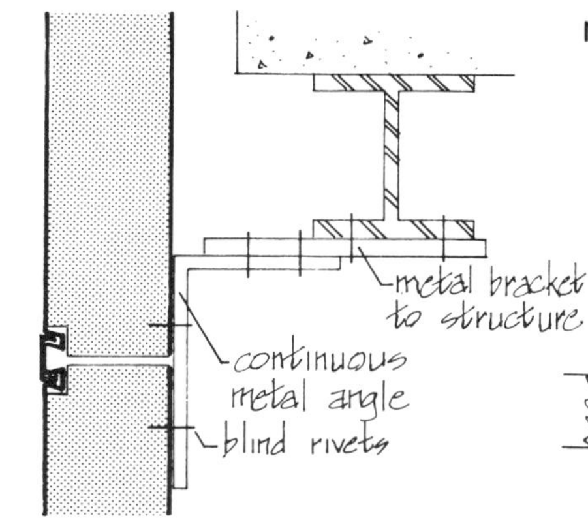

metal frame

gasket

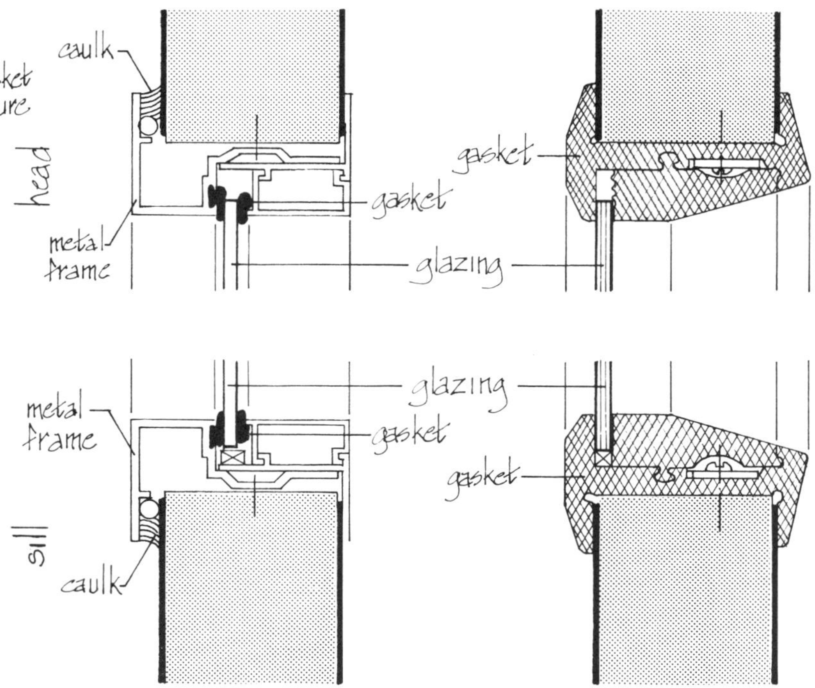

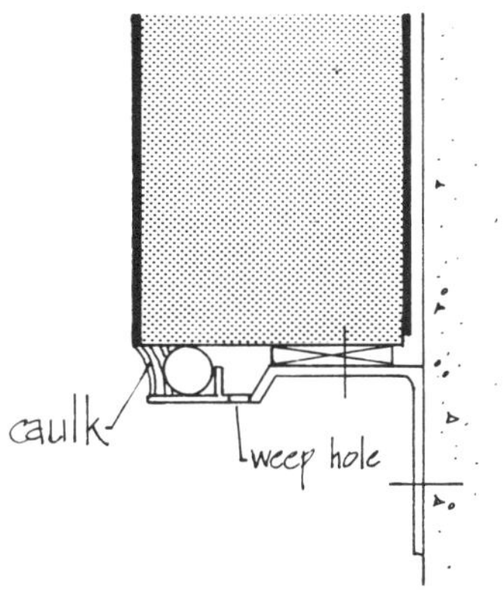

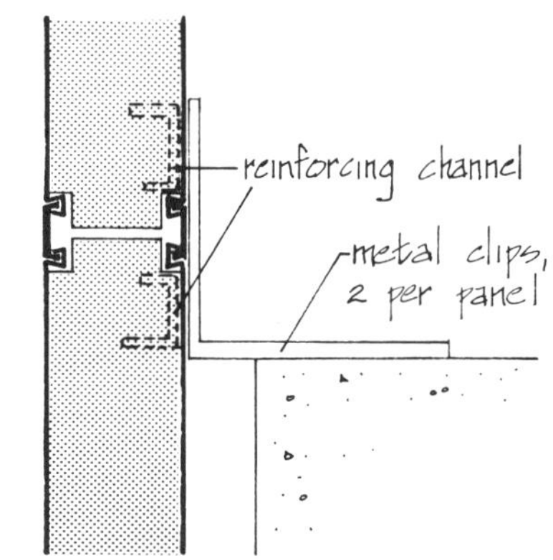

METAL CURTAIN WALL

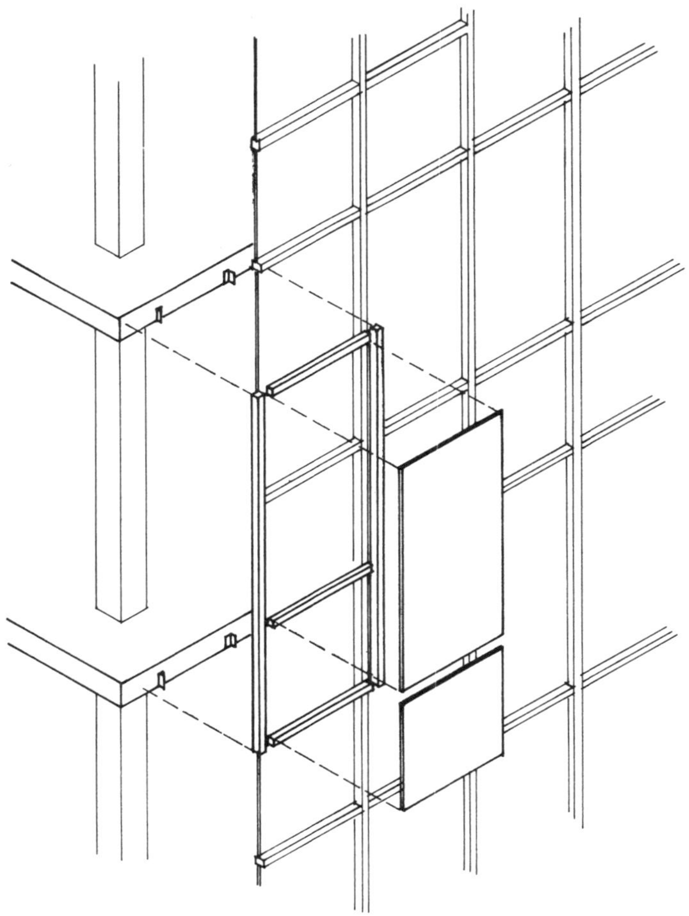

Metal curtain wall systems can be of a variety of forms. The basic definition of curtain wall means that it is a non-load-bearing exterior enclosure system. The metal curtain wall system described here is commonly referred to as the stile and rail type. This is composed of a series of linear, vertical and horizontal, extruded aluminum mullions. These mullions contain grooves for neoprene gaskets and glazing beads which in turn support the glass, metal, or plastic panels.

This form of curtain wall enjoyed much popularity during the 1950s and 1960s. Although still commonly used, with the advent of energy consciousness, the system has been modified. The panels are better insulated, a modification that in a sense turns them into sandwich panels.

A wide range of extruded shapes is available or possible to manufacture. There is a variety of methods by which the curtain wall can be erected, from piece-by-piece assembly on the face of the building to partial assembly before being lifted into place.

This system provides an extremely lightweight, nonstructural building enclosure and its applications have ranged from small one- and two-storey buildings to large high-rise structures.

typical shapes

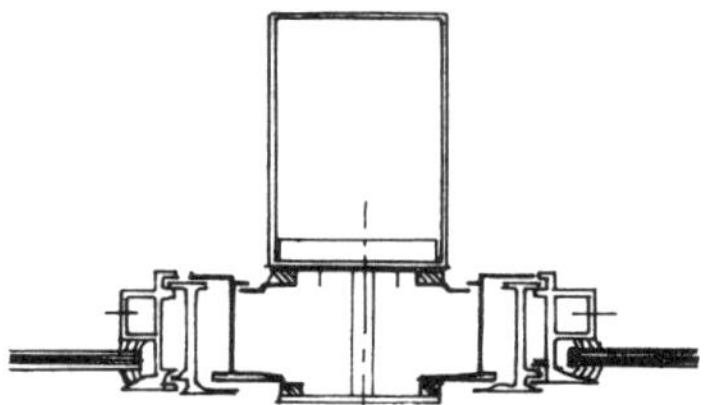

vertical joint closed with batten

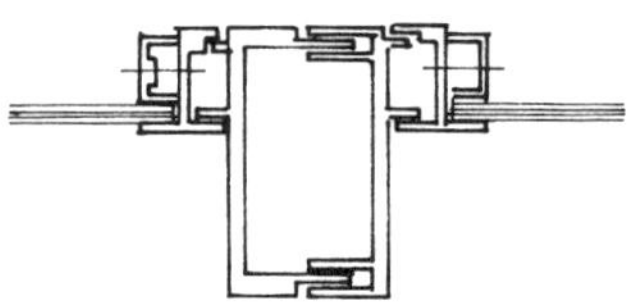

split mullion with mating joint

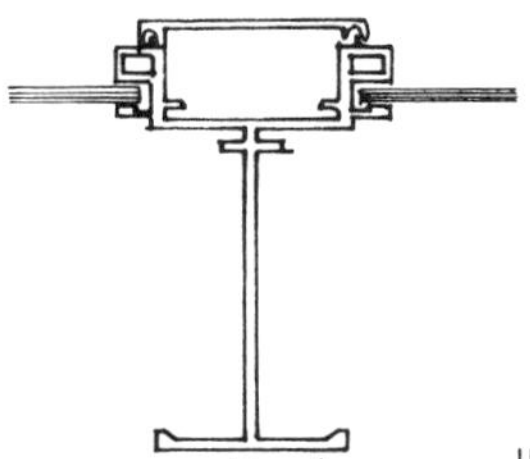

I-shape mullion

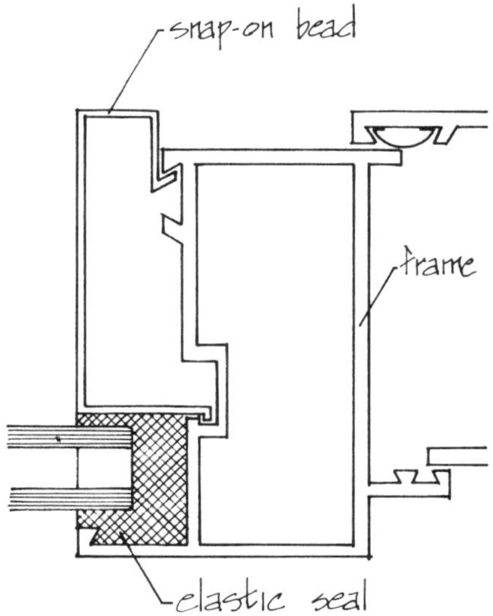

details

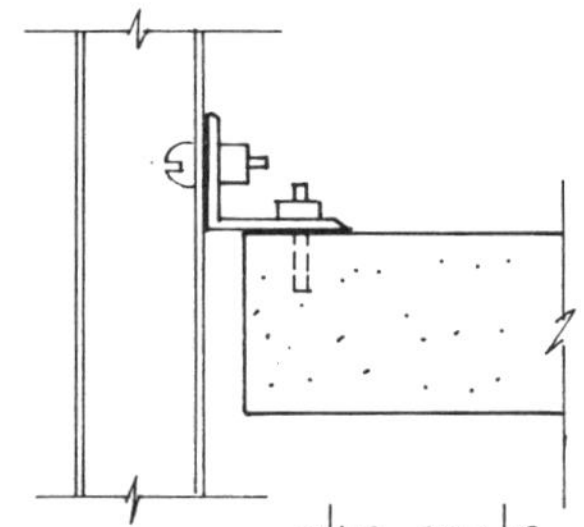

clip angle

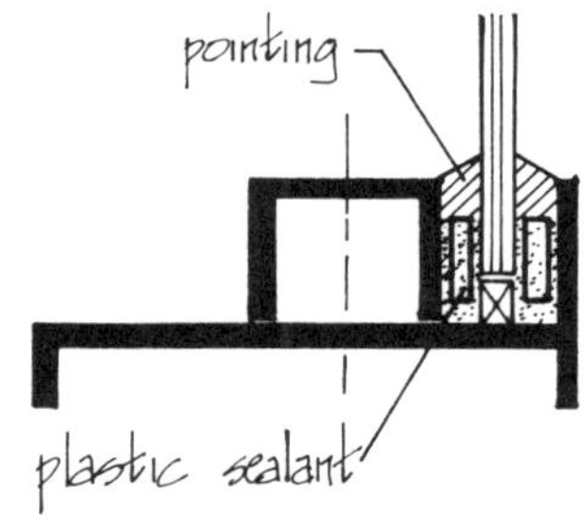

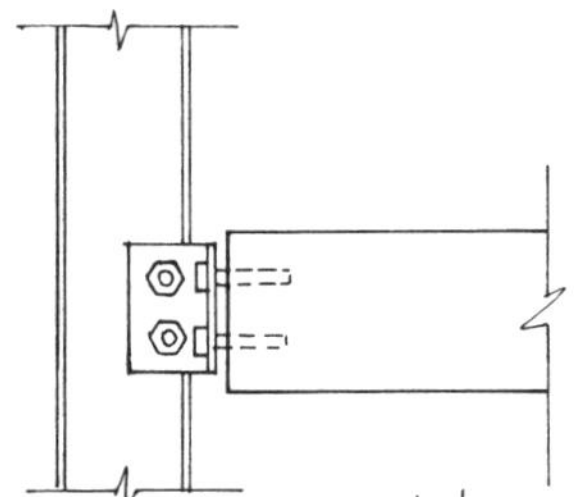

face-mounted angle

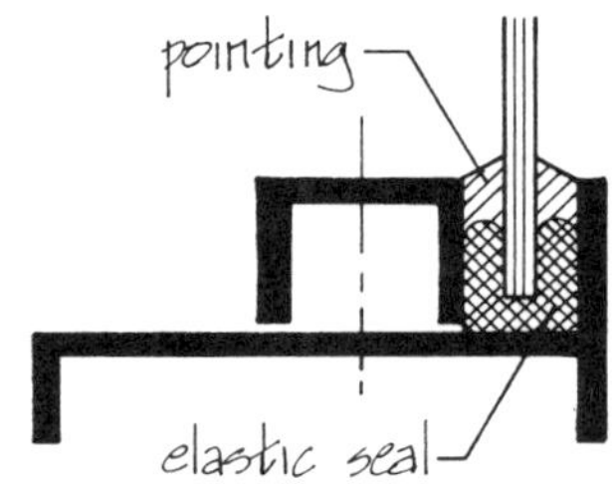

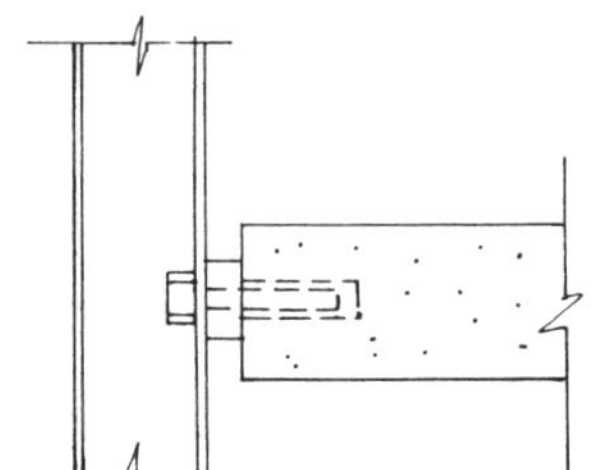

threaded insert

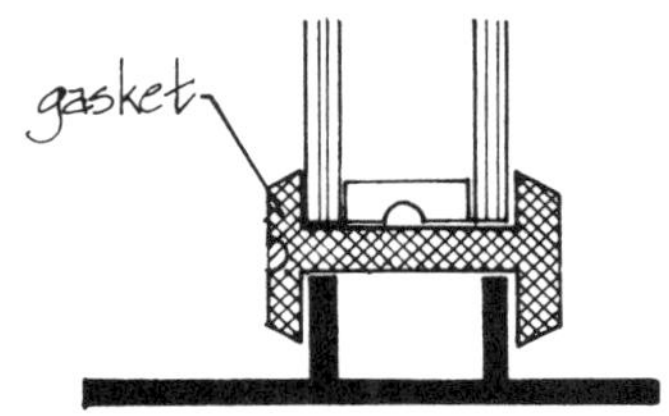

PRECAST PANELS

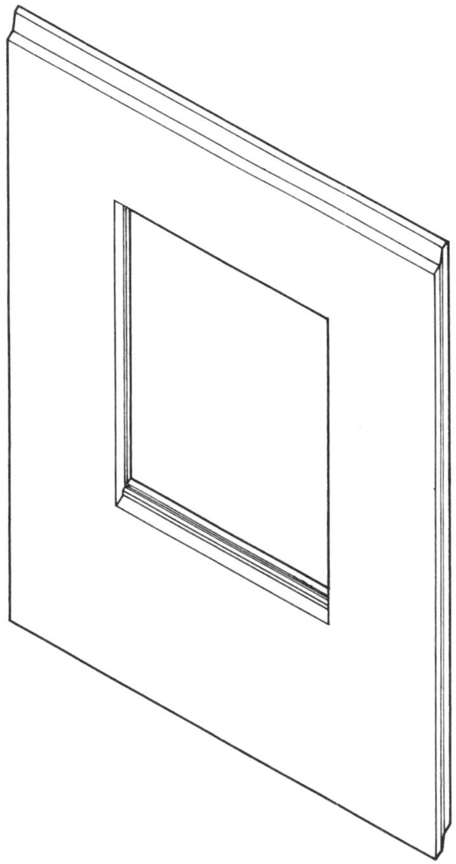

Precast concrete panels can be cast into a variety of configurations, due to the plasticity of the material. Usually they are custom-designed for a specific project, although standard shapes do exist. The most important consideration in its selection is the quantity of similar elements, so that in the precasting operation a standard mold, with minor variations, can be used for all the panels.

Precast panels can be classified into two major categories: flat panels and sculptural panels. Flat panels have a planar, two-dimensional surface; whereas sculptural panels are three-dimensional with projections and recesses. Both types of panels can be either load-bearing or non-load-bearing curtain walls.

The three major characteristics of all precast panels are color, texture, and pattern. Color is dependent upon the type of matrix (cement and aggregate) used. Texture is achieved through an exposed aggregate finish. It can range from a smooth to a coarse surface. The pattern can be created through the formwork or special liners within the formwork. This can range from a high to a low relief.

The common maximum size of transportable units is 20 tons, 8 feet in width, 8 feet in height, and 40 feet in length. Larger sizes are possible with special trailers and permits.

The major advantages in the use of precast panels are greater control over quality and finish, greater strength, and possibilities of prestressing or post-tensioning.

The major limitations in the use of precast panels are moisture protection of joints, transportation, erection, and quantity and repetition of panels needed for economics.

section

details

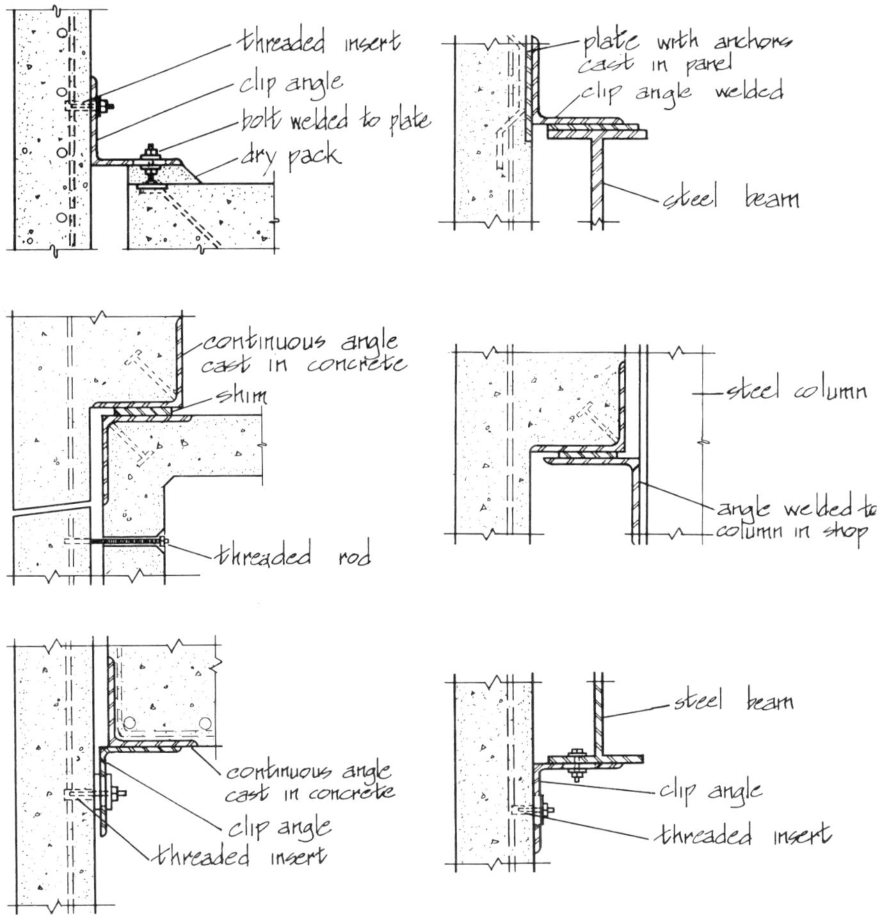

window

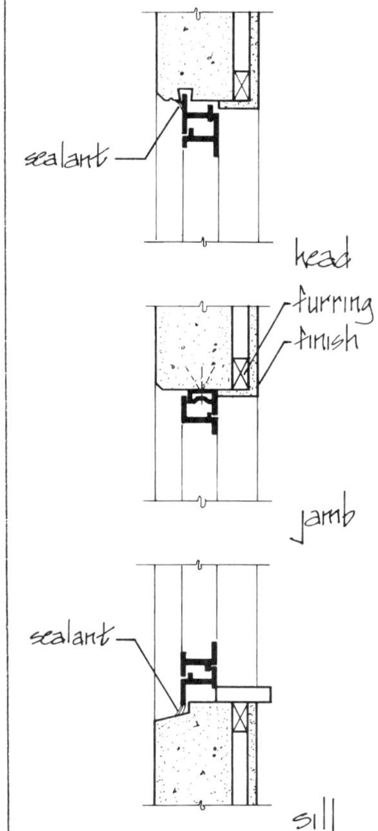

joints

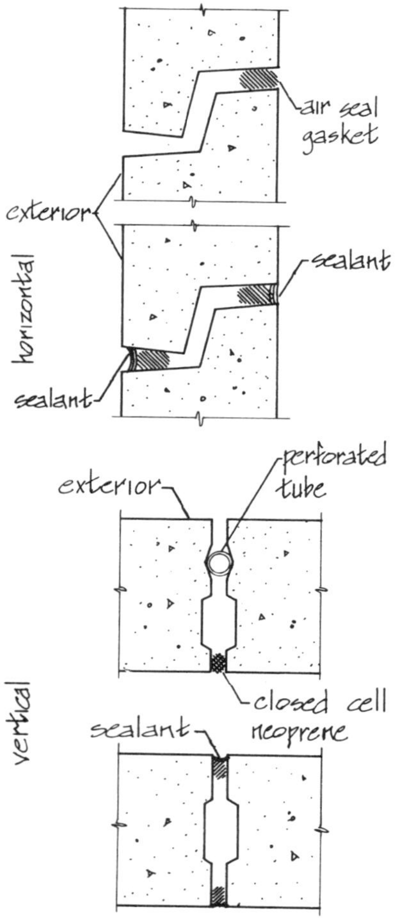

variations

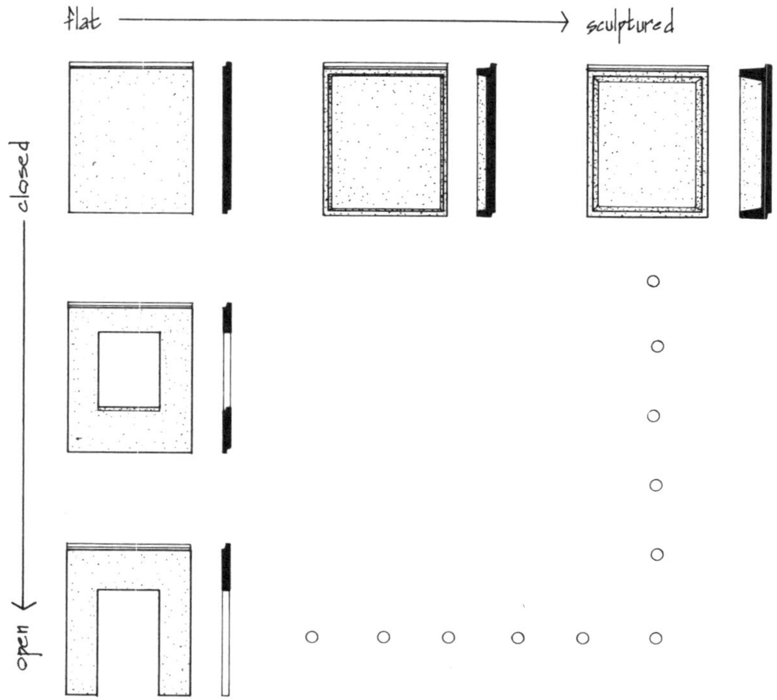

tolerances

Dimensional tolerances of finished units:

10 feet or under	$\pm 1/8$ inch
10 feet to 20 feet	$+1/8$ in., $-3/16$ in.
20 feet to 30 feet	$+1/8$ in., $-1/4$ in.
Each additional 10 feet	$\pm 1/16$ in. / 10 ft.
Angular deviation of plane of side mold	$1/32$ in. / 3 in. depth or $1/16$ in. total, whichever is greater
Opening within one unit	$\pm 1/4$ in.
Out of square	$1/8$ in. / 6 feet or $1/4$ in. total, whichever is greater
Total thickness	$+1/4$ in., $-1/8$ in.

CHART OF MINIMUM PANEL THICKNESSES

Panel Dimension	8′	10′	12′	16′	20′	24′	28′	32′
4′	3″	4″	4″	5″	5″	6″	6″	7″
6′	3″	4″	4″	5″	6″	6″	6″	7″
8′	4″	5″	5″	6″	6″	7″	7″	8″
10′	5″	5″	6″	6″	7″	7″	8″	8″

U VALUES: WALLS, PRESTRESSED TEES, SOLID FLAT SLABS, OR SANDWICH PANELS

Concrete type	Concrete Thickness,* in.	Concrete Resistance, R	Summer $R_{f_o} = 0.25$, $R_{f_i} = 0.68$					Winter $R_{f_o} = 0.17$, $R_{f_i} = 0.68$				
			Insulation Resistance									
			None	2.0	4.0	6.0	8.0	None	2.0	4.0	6.0	8.0
			U	U	U	U	U	U	U	U	U	U
Normal weight	2	0.18	.90	.32	.20	.14	.11	.97	.33	.20	.14	.11
	3	0.27	.84	.31	.19	.14	.11	.89	.32	.20	.14	.11
	4	0.36	.78	.30	.19	.14	.11	.83	.31	.19	.14	.11
	5	0.46	.72	.29	.19	.14	.11	.77	.30	.19	.14	.11
	6	0.55	.68	.29	.18	.13	.11	.72	.29	.19	.14	.11
	8	0.73	.60	.27	.18	.13	.10	.63	.28	.18	.13	.10
Structural lightweight	2	0.47	.72	.29	.19	.14	.11	.76	.30	.19	.14	.11
	3	0.70	.61	.28	.18	.13	.10	.65	.28	.18	.13	.10
	4	0.93	.54	.26	.17	.13	.10	.56	.26	.17	.13	.10
	5	1.16	.48	.24	.16	.12	.10	.50	.25	.17	.12	.10
	6	1.40	.43	.23	.16	.12	.10	.44	.24	.16	.12	.10
	8	1.86	.36	.21	.15	.11	.09	.37	.21	.15	.11	.09

*Thickness of tees is thickness of slab. If walls consist of insulation between two layers of concrete, then thickness shown is the sum of the two concrete thicknesses.

COMPOSITE PANELS

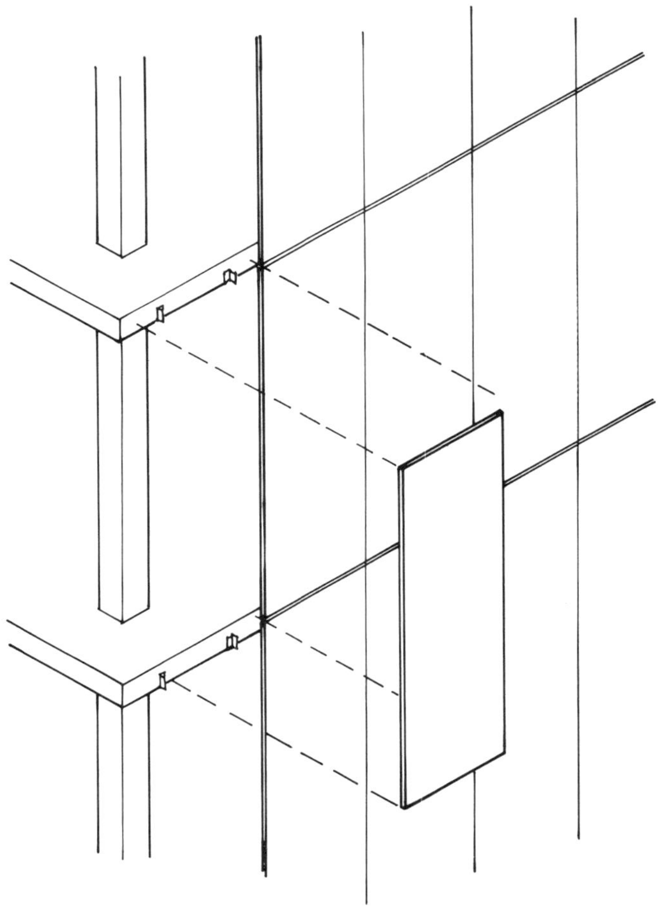

Composite enclosure panels consist of a lightweight cement asbestos board core, which may then have a variety of aggregate and textured finishes applied. They are used as nonstructural, curtain-wall enclosure panels.

The panels can be fabricated in a variety of sizes, with maximum dimensions generally 4 feet by 12 feet. The thickness of the panel can range from 1/8 inch to 1/2 inch. An expanded polystyrene core or other insulating material is a possible option.

Attachment of the panels in the field is generally by welding or bolting the panels to a frame of steel studs or similar support surface.

Due to the size of the panels, a wide range of configurations and applications is possible. They can be used for practically any building application, both new construction and renovation.

types & sizes

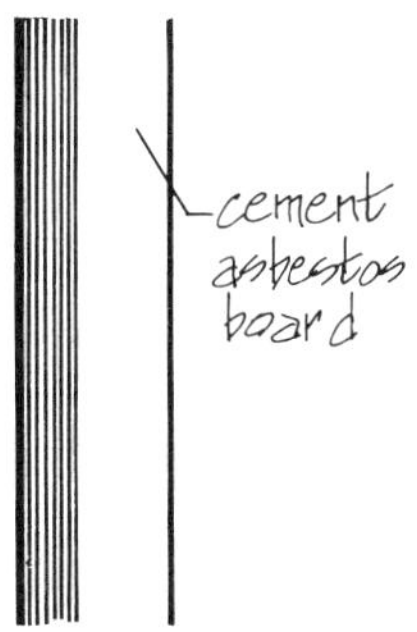

uninsulated panel

thickness	2'×4'	4'×4'	4'×8'
1/8"	35	15	10
3/16"	65	45	25
1/4"	90	65	40
3/8"	160	115	75
1/2"	210	150	100

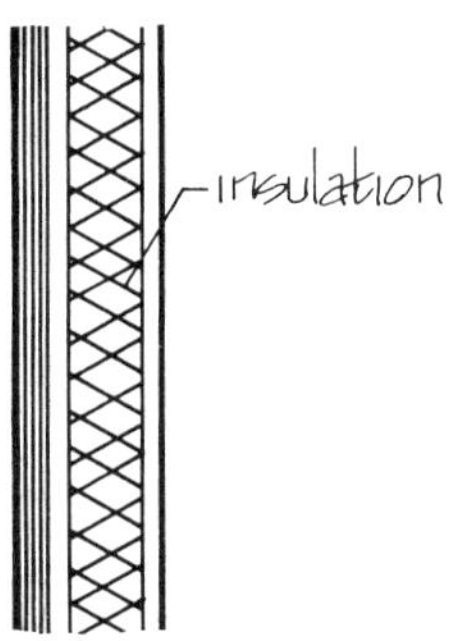

insulated panel

thickness	"U" factor	3' spacing	4' spacing
1"	.25	34	24
1 1/4"	.19	46	32
1 1/2"	.15	58	40
1 3/4"	.13	70	48
2"	.12	82	56
2 1/4"	.10	90	64

details

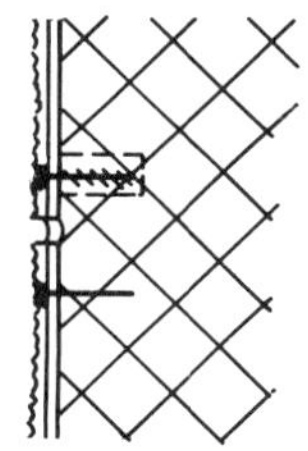
masonry

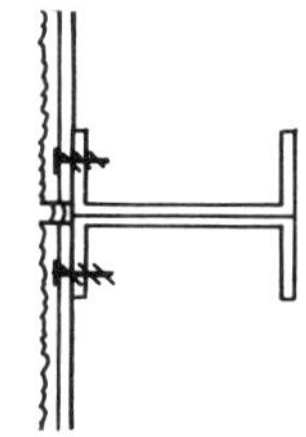
metal frame

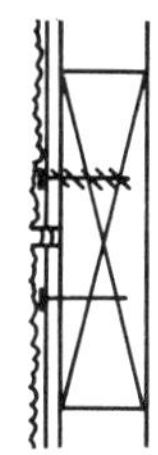
wood

metal girt

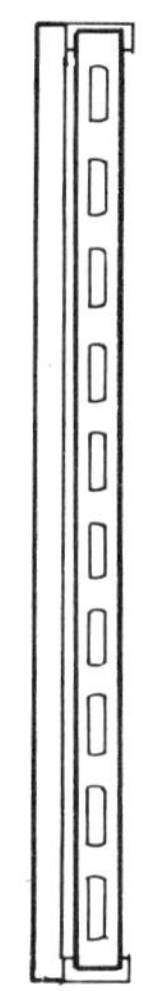
combined stud & panel

COMPONENTS

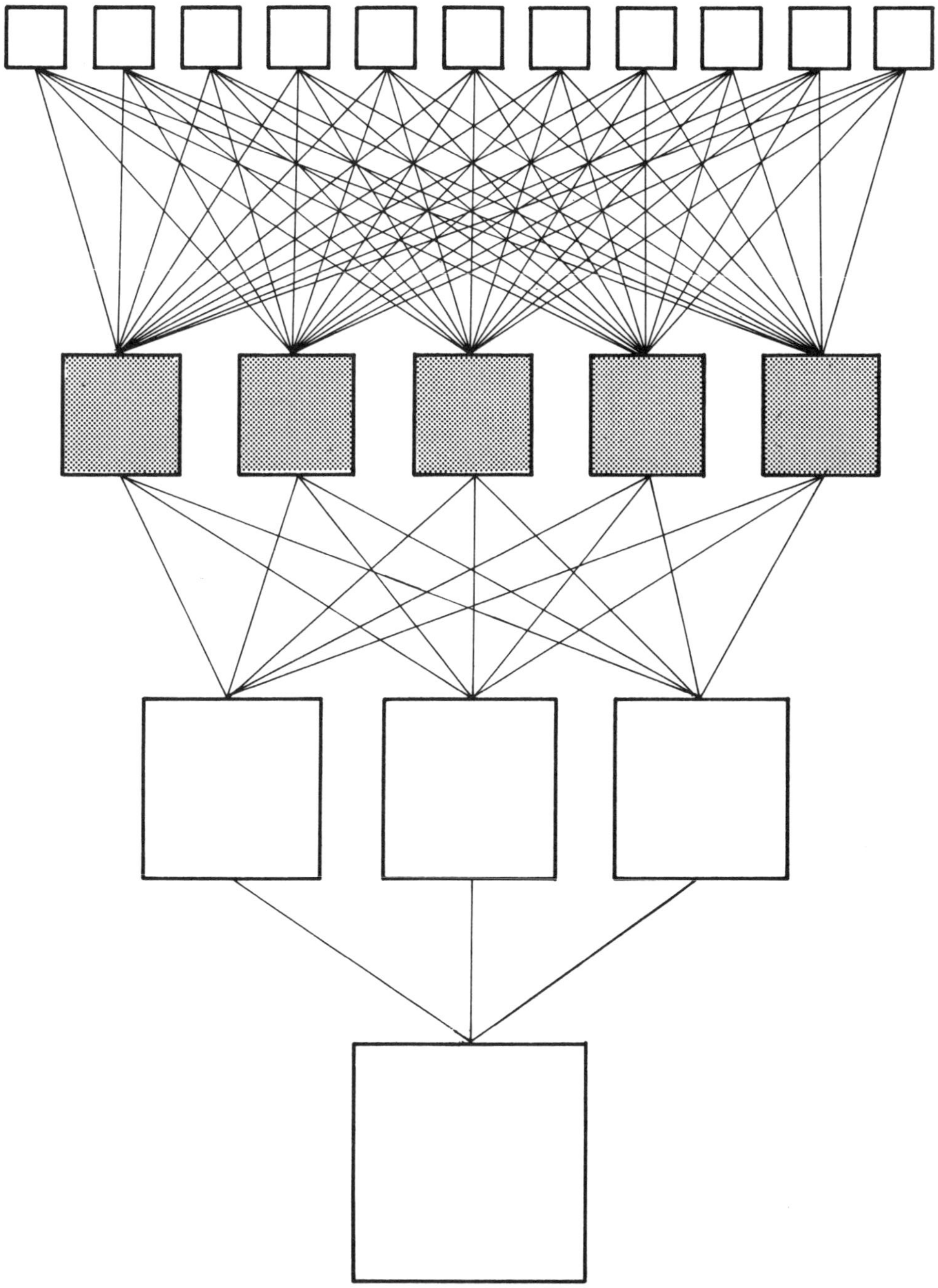

STEEL JOISTS/OPEN WEB

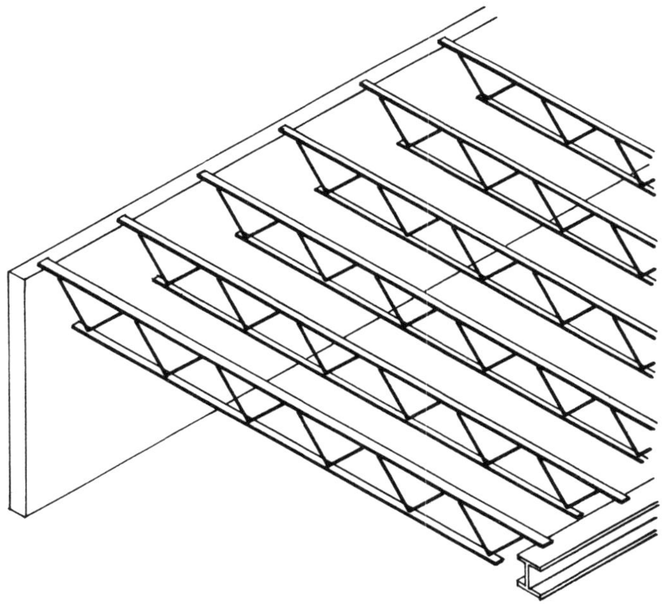

Steel bar joists are standardized lightweight metal trusses which can be used to span large areas and accommodate a variety of loading conditions. Supported by beams or bearing walls, they may be used for either floor or roof construction. The standard configuration is a flat truss; however, the top or bottom chords may also be pitched. Horizontal or diagonal bracing is required to prevent lateral movement of top and bottom chords. The spacing of bridging depends on chord size and joist span, and varies from 10 feet to 15 feet on center. As a general guideline, the span of steel joists should not exceed 24 times the joist depth.

The interface with finishing materials can be one of several options. A ceiling may be hung directly on, or supported from the bottom chord. Mechanical lines may be run through the openings in the truss, or may be hung below. The applied decking may be:

1. composite steel deck and reinforced concrete slab
2. precast concrete plank
3. cast-in-place reinforced concrete slab
4. wood plank

Applications of open web steel joists have included commercial, office, and industrial construction. The most severe limitation is the depth required, but this is offset by the large spans and light weight possible.

types

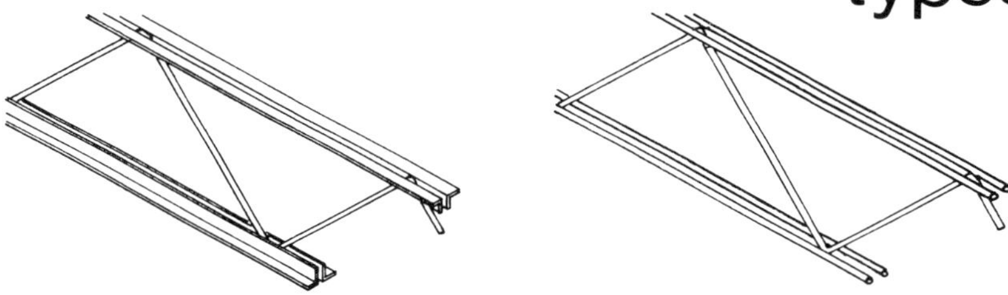

TYPICAL STD. SIZES & SPANS

joist series	depth	max. span
J & H	8" to 30"	60'
LJ & LH	18" to 48"	96'
DLJ & DLH	52" to 72"	144'

details

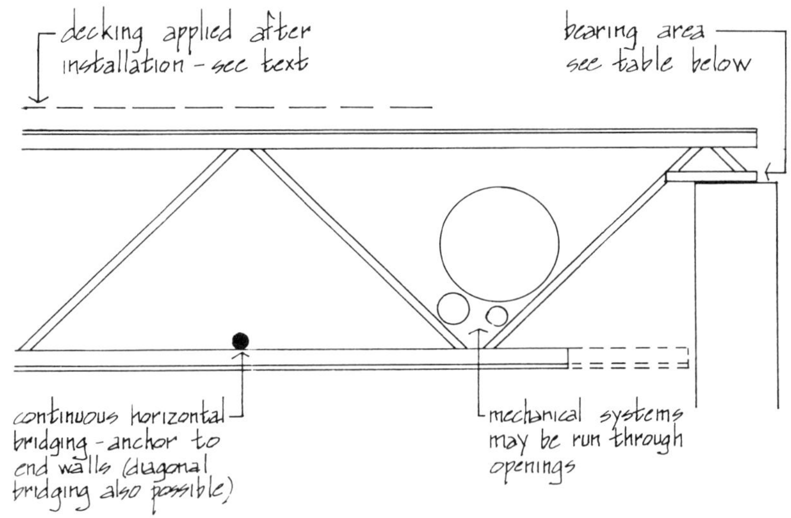

MINIMUM BEARING AREAS

joist series	support condition		
	steel beam	concrete wall	masonry wall
J & H	2½"	4"	4"-6"
LJ & LH	4"	6"-9"	6"-12"
DLJ & DLH	4"	6"-9"	6"-12"

SPACE FRAMES

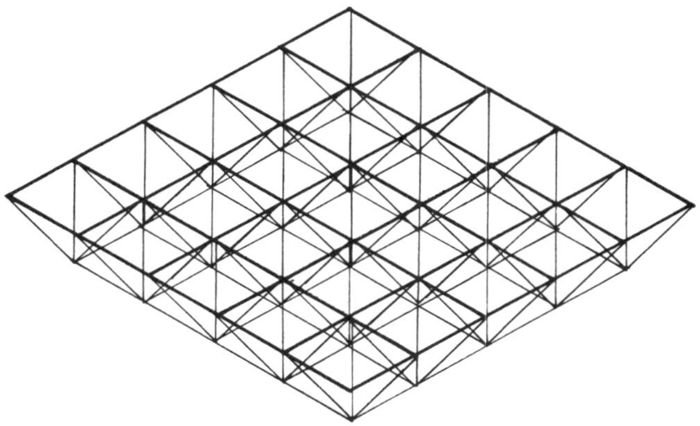

Space frames are manufactured from metal wide flange, tee, tube or pipe sections. These sections are connected to produce a three-dimensional truss system capable of spanning large distances with a minimum of support points.

The space frame is composed of subsections, or modules, which may vary in size due to design and application. Some common sizes for these modules are 4 feet, 5 feet, 11 feet, and 17 feet. Orthogonal or skewed space frame configurations are possible. In the orthogonal space frame, the truss chords are parallel to the perimeter of the total space frame. A skewed configuration has the top and bottom chords oriented at an angle, usually 45°, to the perimeter.

The fabrication process usually involves production of shop-welded subassembly units, which are then shipped to the site for erection. At the site these subassemblies can be erected as self-supporting units which are bolted to adjacent units. An alternate method involves connecting the subassemblies on the ground and then installing the space frame as a single unit.

Finishes and interfaces with other systems are easily handled. The standard roof deck consists of prepainted ribbed metal decking which is tack welded to the top chord. Acoustical treatment, insulation, and final roofing material may then be applied. Skylights can easily be incorporated within this system. The space between the top and bottom chords can be utilized as a location where lighting, HVAC, and other equipment may be placed.

Advantages of space frames lie in their light weight, large span, and aesthetic appeal.

configurations

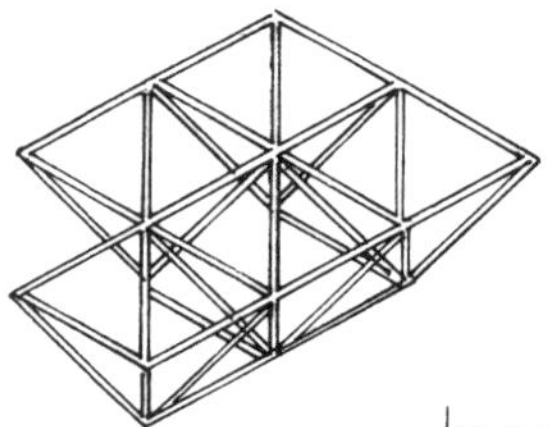

basic modules

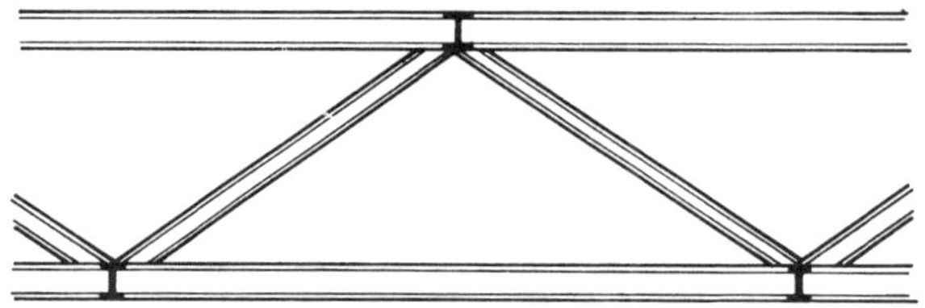

flange

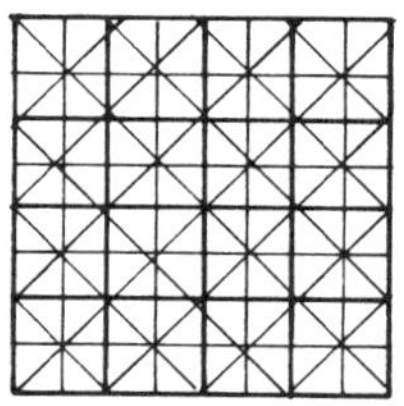

rectangular grid

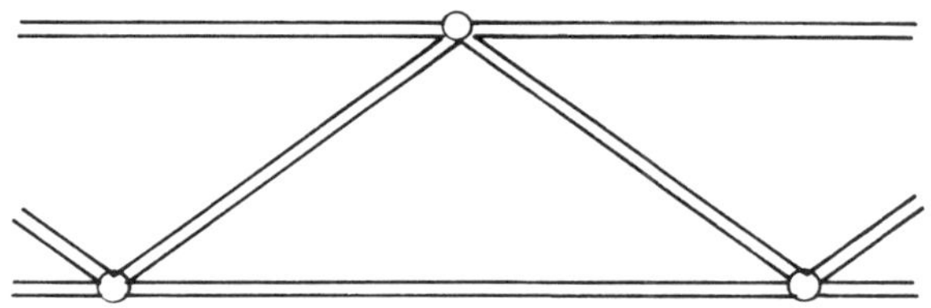

tubular

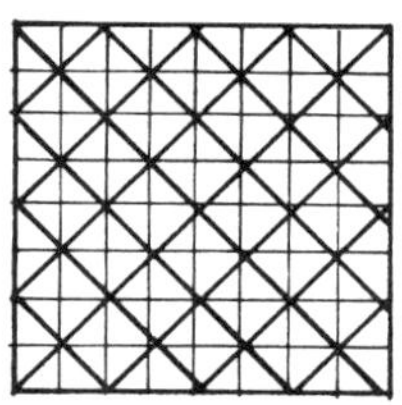

skew grid

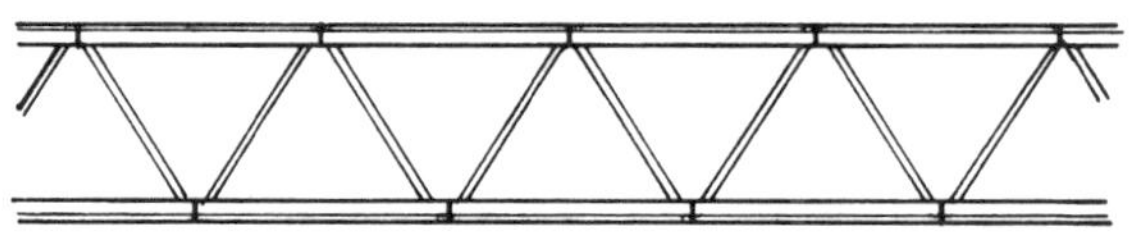

angles

spans

module	spans
4'-0"	28' to 144'
5'-0"	30' to 145'
11'-3¾"	100' to 240'
16'-11⅝"	168' to 324'

details

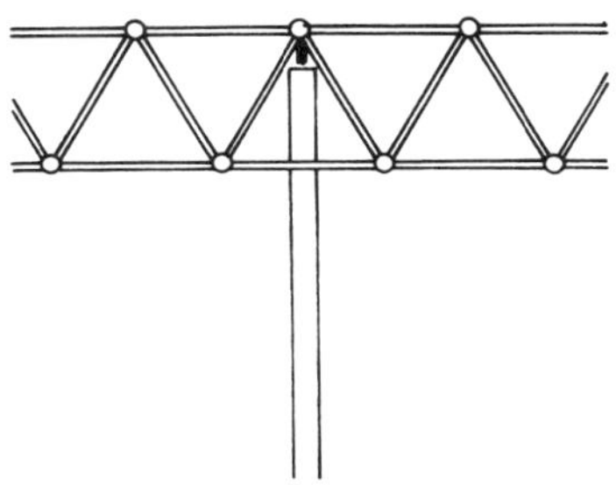

internal column

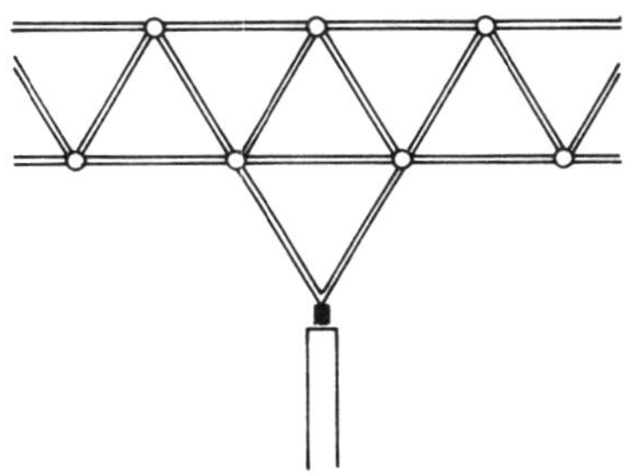

dropped module

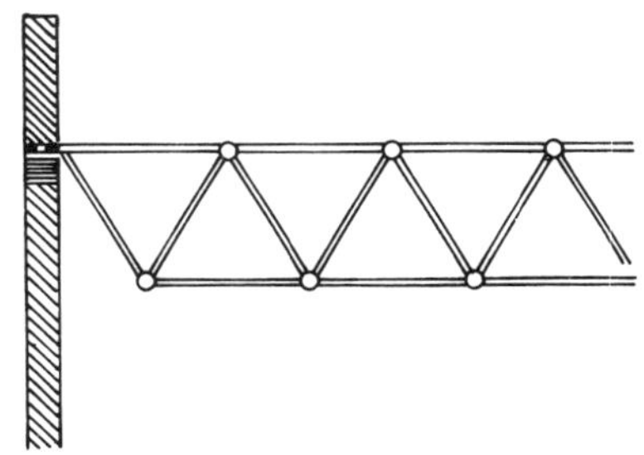

masonry wall (end condition)

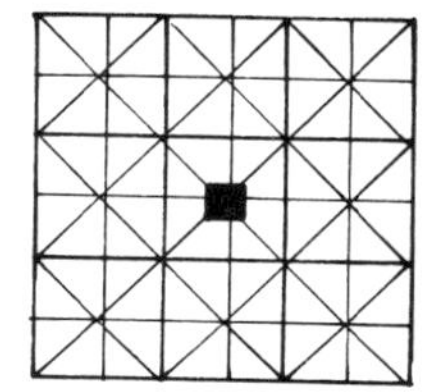

column support

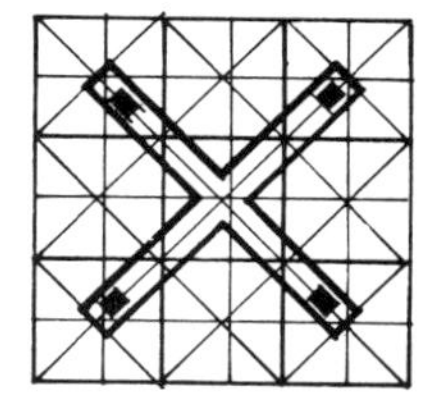

cruciform column support

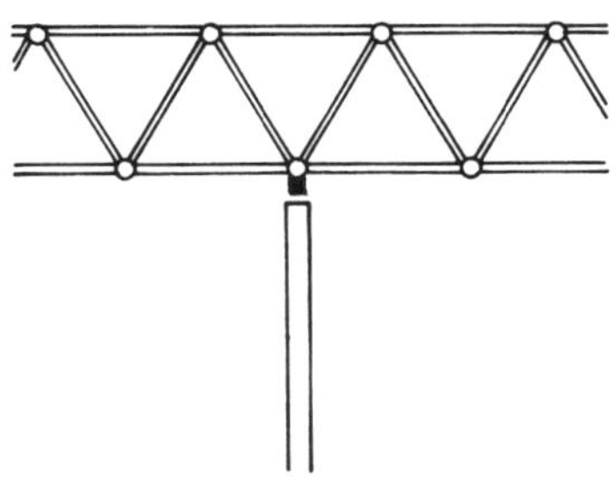

lower chord support

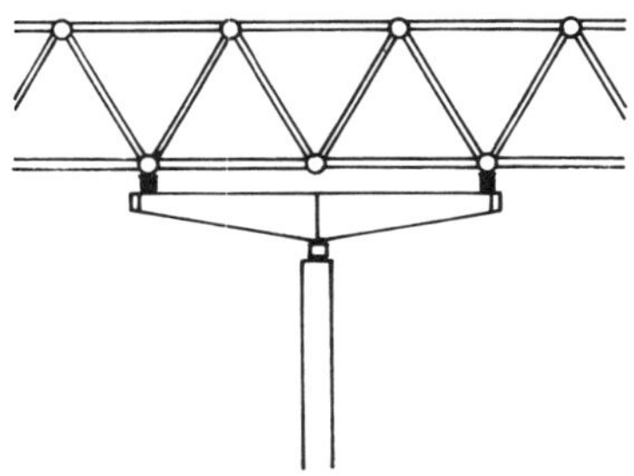

cruciform column

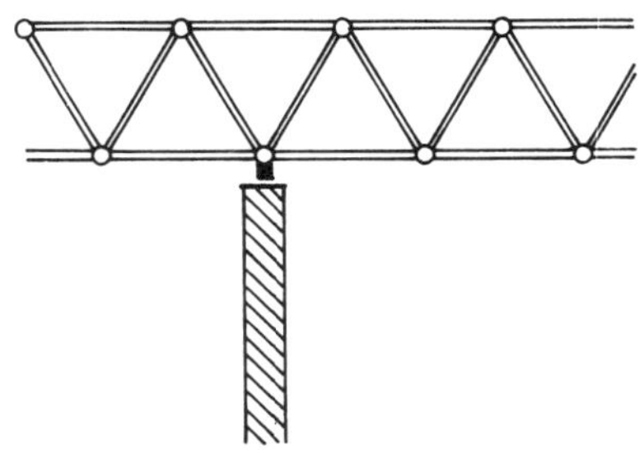

masonry wall (internal)

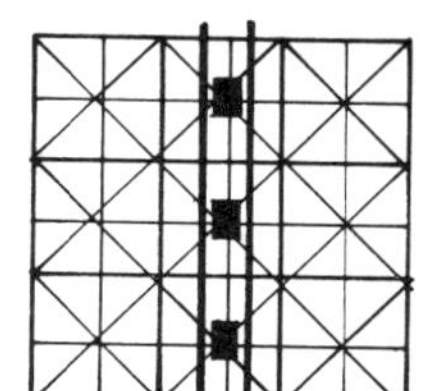

bearing wall support

ceiling details

lighting / ceiling

partition

services

air supply & return

PRECAST CONCRETE MEMBERS

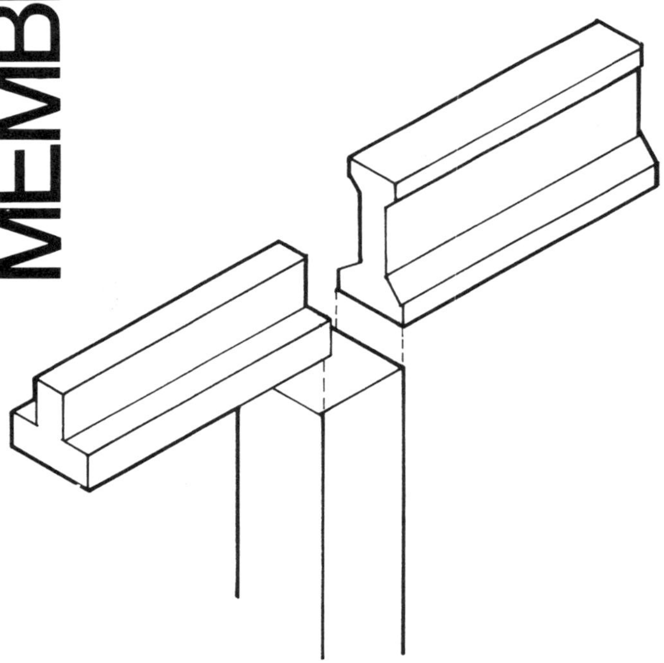

Precast concrete members are available in a wide range of standard structural shapes — beams, columns, and joists. Many are readily available in a variety of sizes from manufacturers. Special details, sizes, or configurations can also be produced. They can be used with other materials such as masonry, or combined to form a totally precast structural frame.

Precast members have higher strength than poured-in-place concrete, and this can be increased further by the use of pre-stressing or post-tensioning. Precast concrete allows greater uniformity of the material due to factory production and quality control. It also retains the general characteristics of the concrete material — inherent fire resistance, low maintenance, moldability, and low cost.

typical shapes

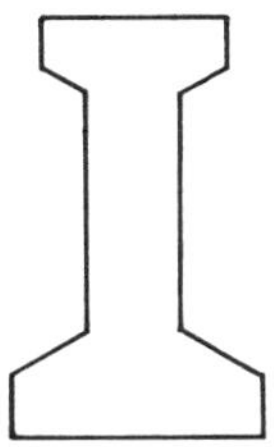

I-beam

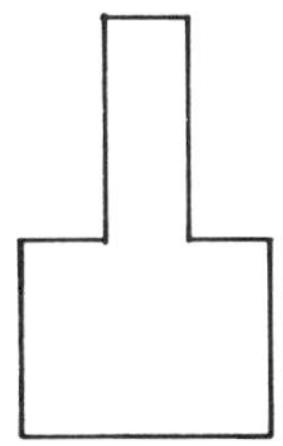

inverted tee-beam

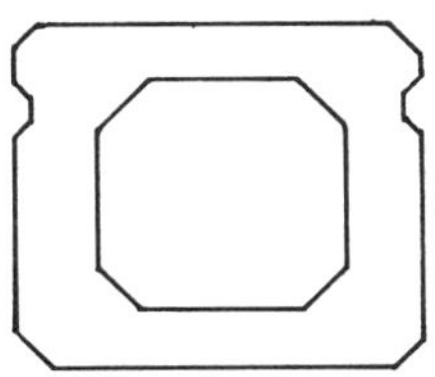

box beam

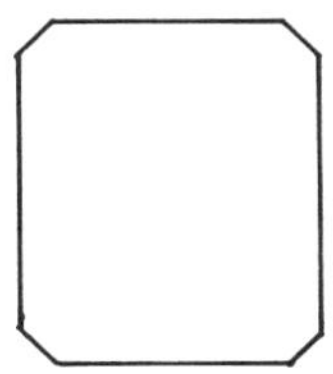

solid column

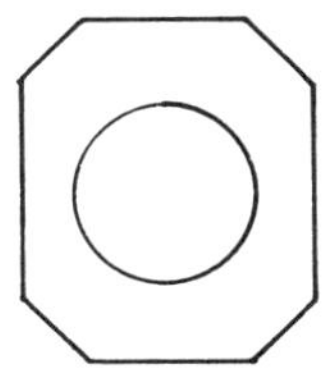

hollow column

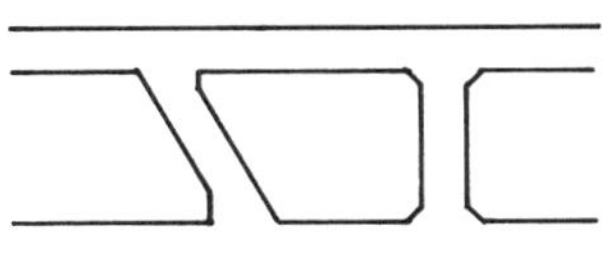

truss

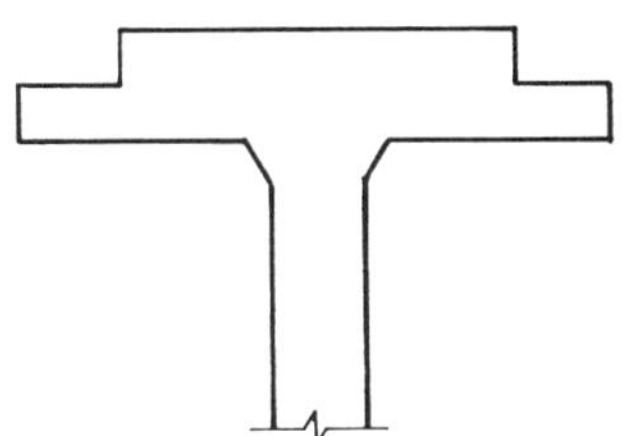

T-column

connections

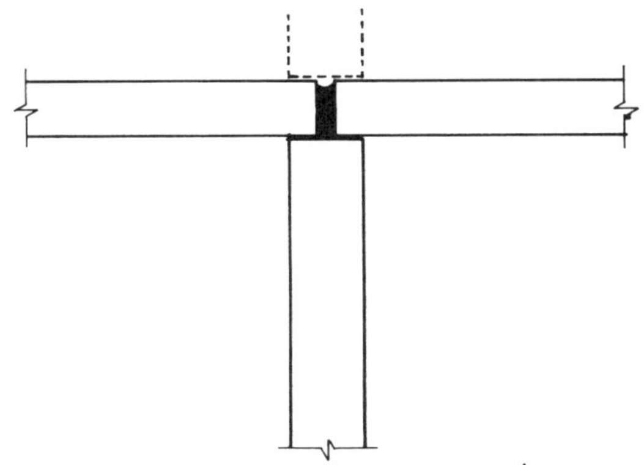

column & beam

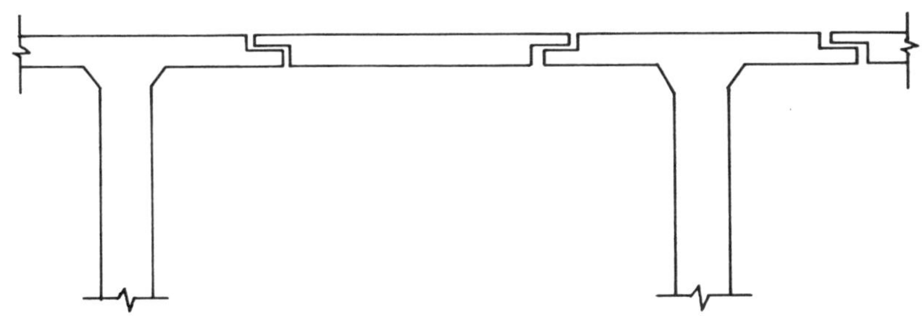

T-columns with beam

PRECAST TEES

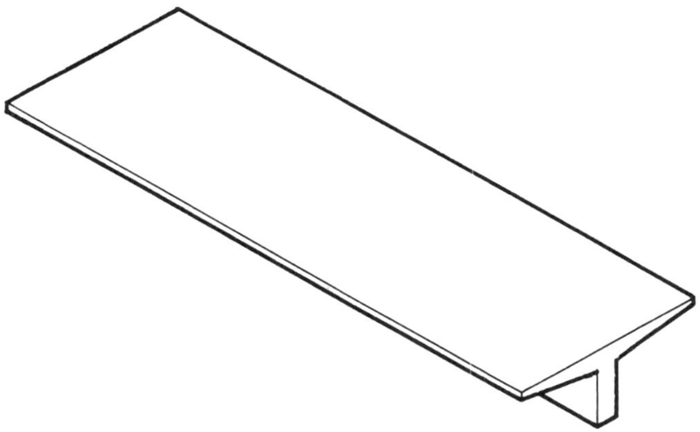

Precast concrete tees and double-tees are commonly used precast elements. They are generally used for roofs, but they can also be installed vertically to serve as wall panels. They are essentially a thin slab with a stiffening rib or ribs, which acts like beams. In this way, extremely strong elements are produced which are capable of spanning large distances with a minimum of material.

Although single- and double-tees are the most common shapes, other configurations are possible such as vee-shaped or gull-wing. Used vertically, openings for doors and windows can easily be provided. Connection of tees to beams is generally by means of a welded connection, and to columns is by means of a dowel.

sizes — configurations — details

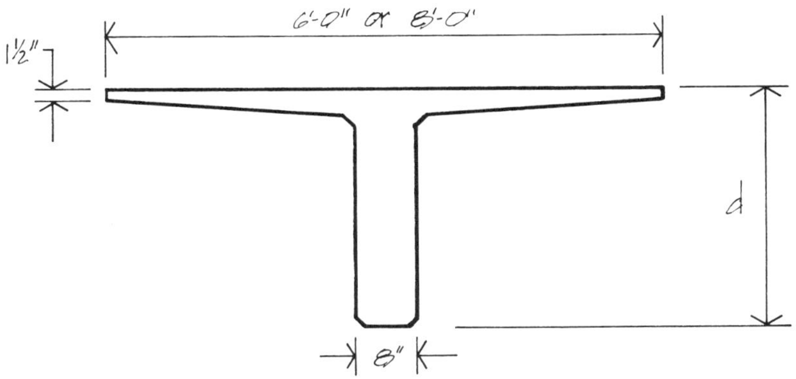

typical maximum spans

d	unit width	
	6'-0"	8'-0"
28"	80'-0"	72'-0"
32"	96'-0"	86'-0"
36"	112'-0"	98'-0"
40"		108'-0"

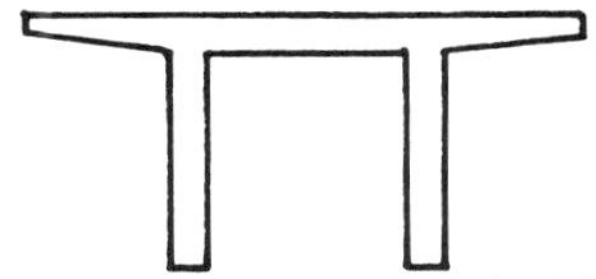

double tee

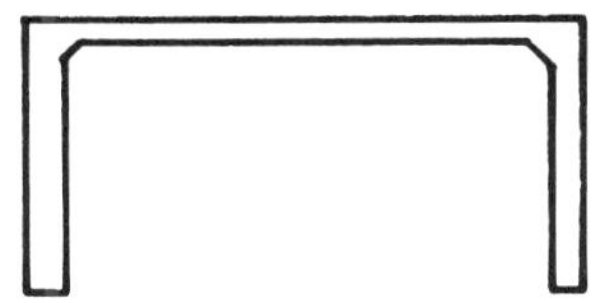

inverted channel

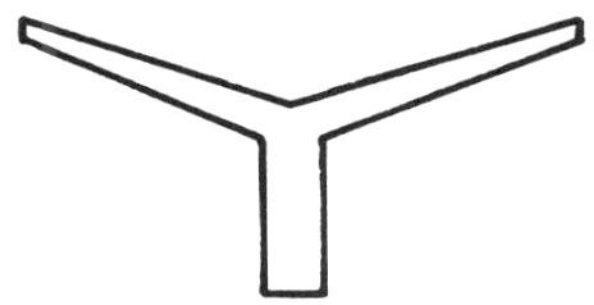

Y-shape

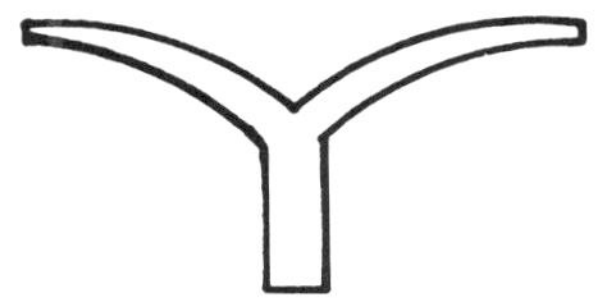

gull-wing

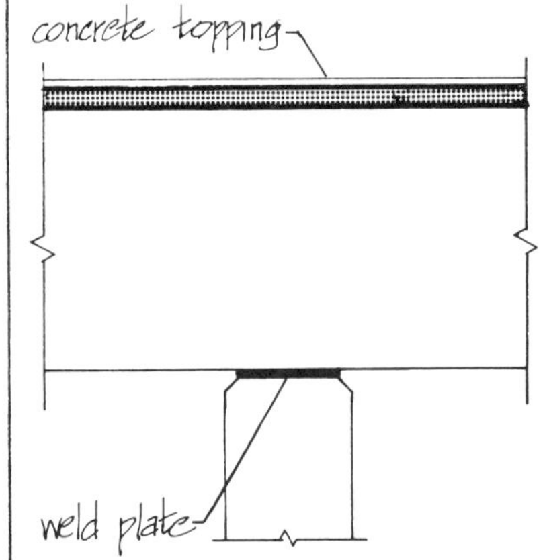

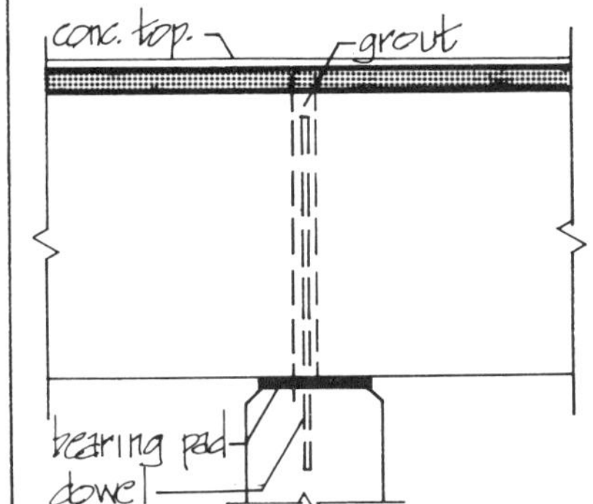

CONCRETE PLANKS

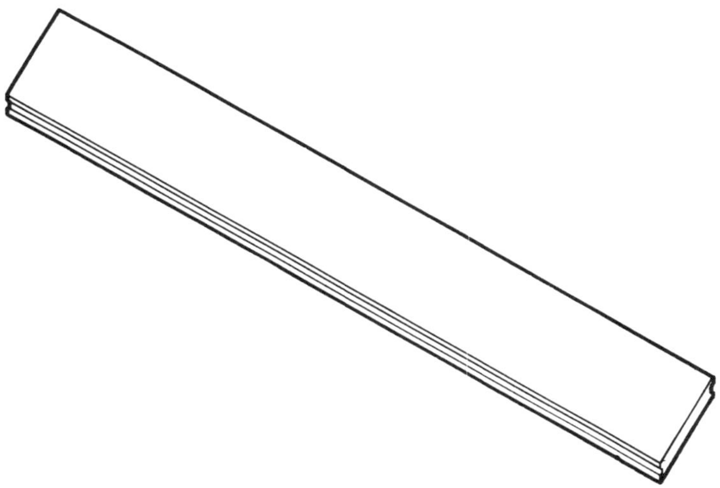

Concrete planks are hollow, precast pre-stressed slab units used for floor and roof applications. They consist of rectangular units 4 feet wide and 6 to 12 inches in depth. They can span up to 48 feet without intermediate support. For special applications the planks can be notched, beveled, cast with weld plates, or cast in special widths.

The planks are compatible with practically any type of framing system: precast bearing walls, masonry bearing walls, precast concrete frame, steel frame, and cast-in-place bearing walls. Cantilevers for balconies or overhangs are easily accomplished. The individual slabs are joined by grouting to form a flat, rigid deck. A concrete topping slab may also be poured to connect them.

The hollow cells can be used as air ducts for heating and cooling and as electrical raceways. The units have satisfactory resistance to sound transmission, heat conductance, and fire. The floor finish may be any of the range of finishes normally possible over concrete decks. The smooth under-surface of the slabs may be utilized as a finished ceiling by simply caulking and painting. Other ceiling finishes are easily applied as well.

The adaptability of precast planks permits their use in a variety of types of low-, medium-, and high-rise buildings.

details

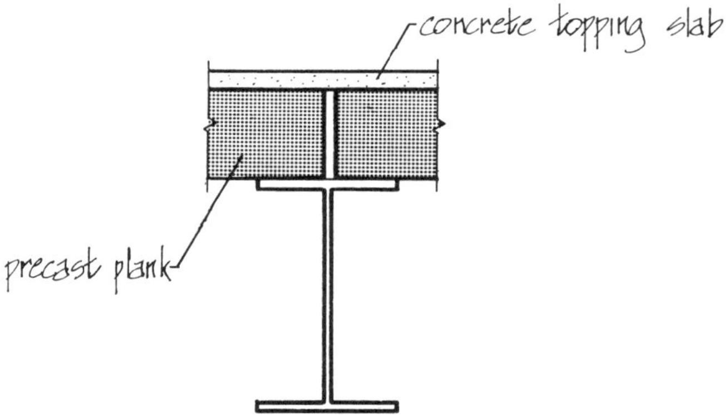

steel beam

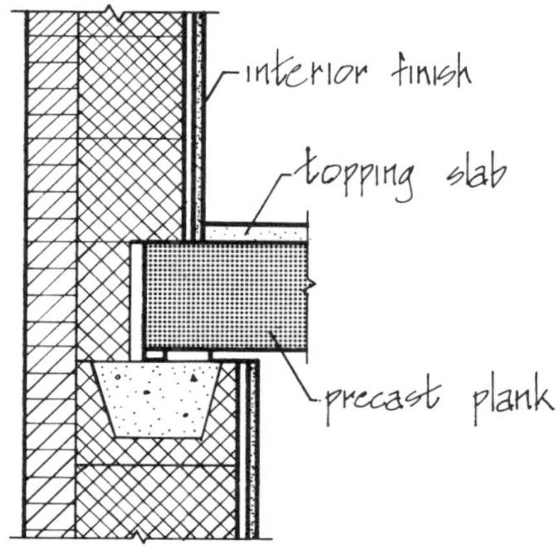

masonry wall

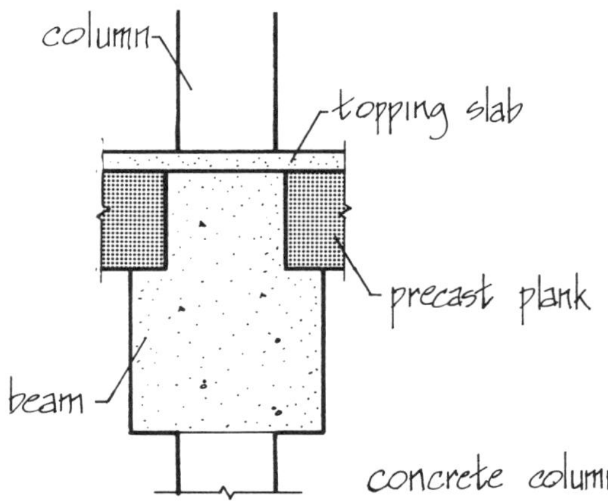

concrete column & beam

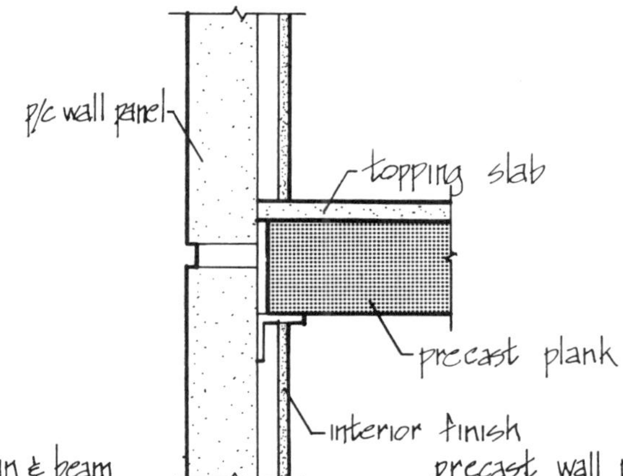

precast wall panels

sizes

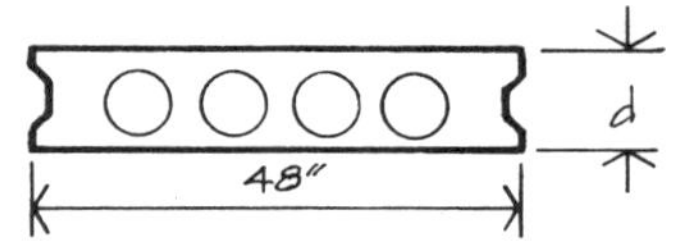

d	max. span
4"	16'-0"
6"	20'-0"
8"	26'-0"
10"	34'-0"
12"	38'-0"

LAMINATED TIMBER

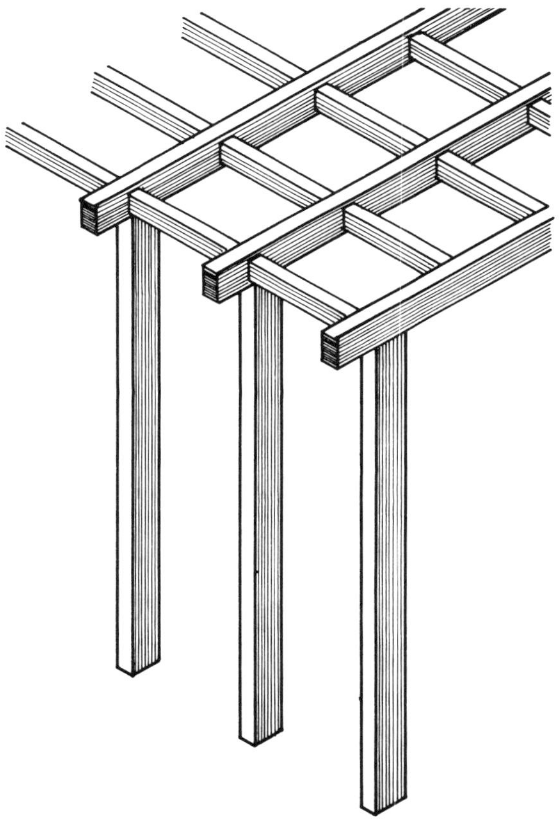

Laminated timber structural elements are produced by bonding together with a waterproof adhesive, pieces of structural grade lumber to produce sizes and shapes difficult to obtain in solid lumber. Columns, straight and tapered beams, purlins, and radial and tudor arches are some of the more commonly produced elements.

Laminated beams are normally spaced 4 to 24 feet on center. Curved arches are normally designed so that the rise is approximately one-fourth the span. More detailed design data is available for precise engineering calculations.

The finishes on laminated timber elements are classified according to three appearance grades — Premium, Architectural, and Industrial. They can be left with a natural finish, stained, or painted to accommodate the specific application.

A wide range of connections to other structural and framing elements is possible. The laminated timber elements arrive at the job site cut to exact size and precisely drilled, thereby simplifying their assembly and erection.

shapes

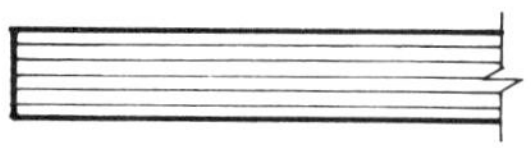
flat beam

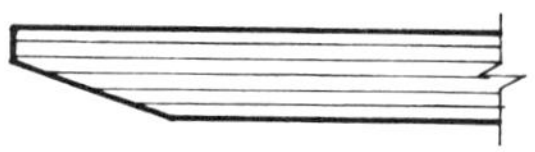
tapered beam

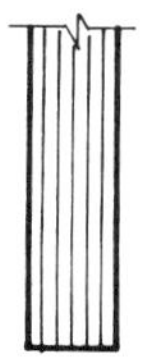
straight column

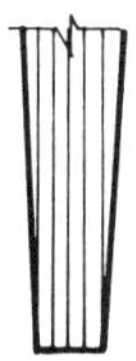
tapered column

details

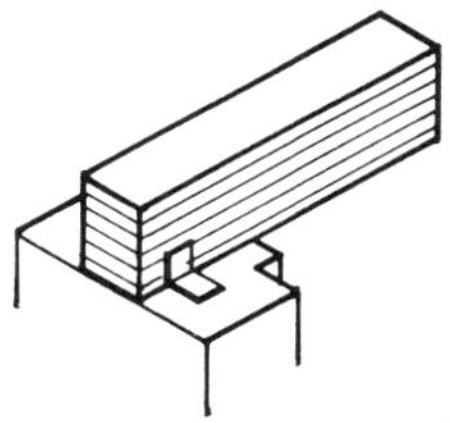
masonry wall

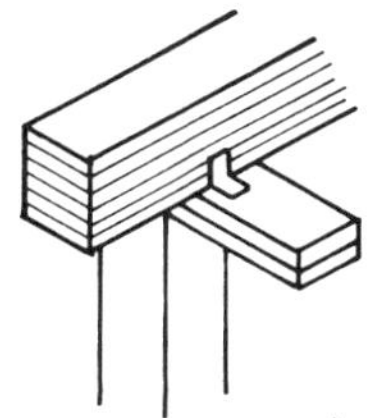
frame wall

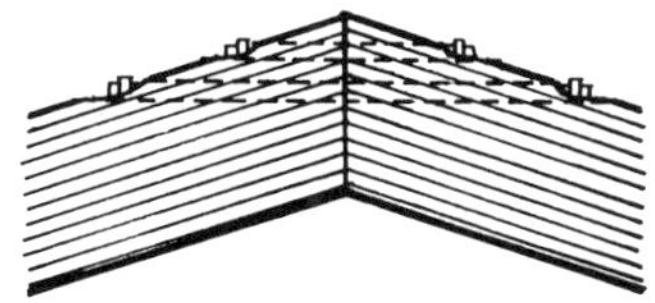
recessed bolts

laminated column

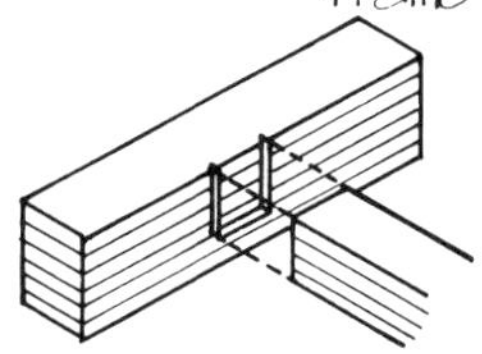
joist hanger

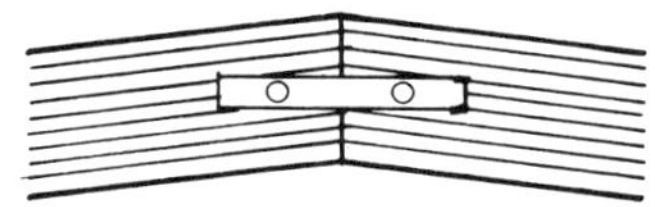
metal strap

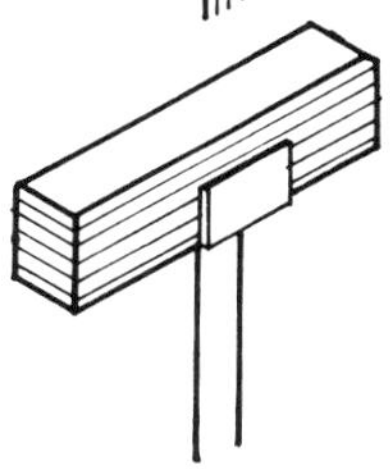
steel column

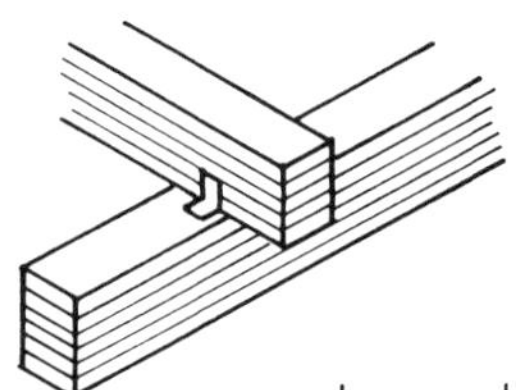
clip angles

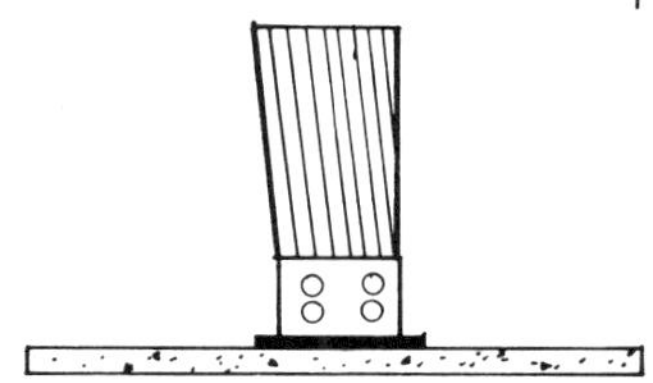
base plate

WOOD TRUSSES

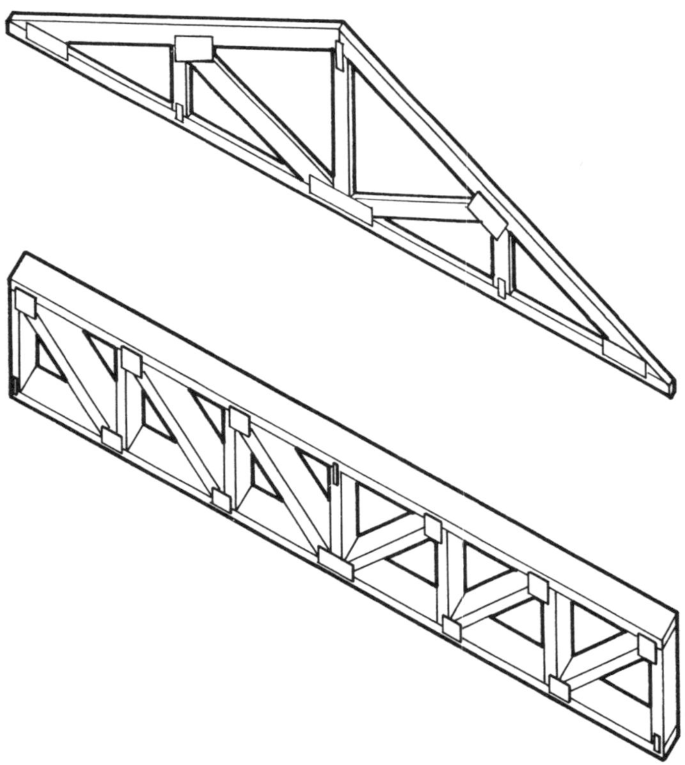

Prefabricated wood trusses represent a first generation stage of industrialization. These elements are constructed in fairly identical fashion as conventional on-site trusses, using a conventional material — wood. The advantages obtained by industrialization are in the areas of job-site time and labor savings, improved quality control, and scheduling. Standard configurations are readily available, and special designs can be produced by the manufacturer.

The manufacturing process involves assembling pre-cut, dimensionally stable, stress-graded lumber on a jig and attaching special connector plates to join the elements into the finished truss. The completed trusses are then either stored at the manufacturing site or transported to the job site.

At the job site, the trusses can be erected by a small, single-trade crew of two or three men. Since in most cases, the truss weighs in the range of 100 to 200 pounds, they can be put in place by the crew without any machinery or with only a small portable winch or crane. The savings in construction time can be approximately 30 percent.

Since the industrialization of this element retains the traditional materials and configurations, the interface with the other elements of the building remain fairly conventional — wood frame and masonry being most common. Mechanical and electrical systems may be incorporated easily by utilizing the spaces between the members of the truss and avoiding notching the members or dropping the ceiling.

Application of finishes is accomplished in the traditional manner. A one-hour fire rating can be obtained by the application of appropriate ceiling materials. Wood trusses are generally used in building types such as low-rise single- and multiple-family housing, small commercial and industrial buildings, and farm structures.

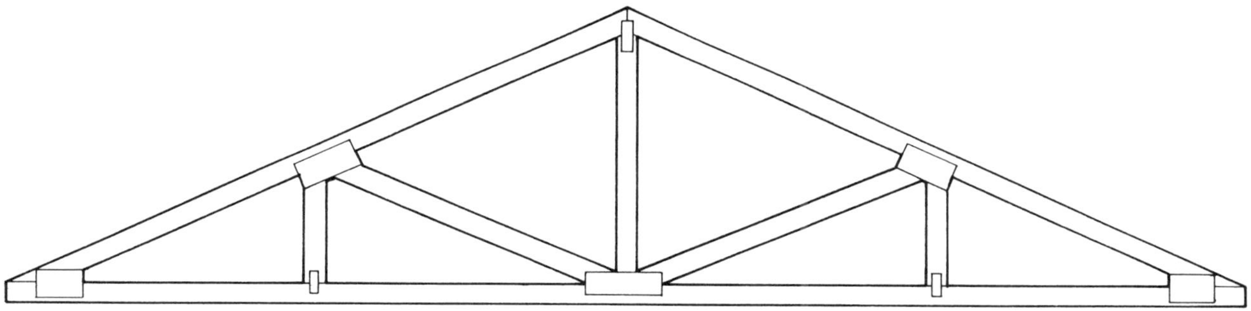

howe truss

chord size		pitch				
tc	bc	2/12	3/12	4/12	5/12	6/12
2×4	2×4	25'	33'	37'	39'	41'
2×6	2×4	25'	34'	42'	48'	53'
2×6	2×6	39'	50'	55'	59'	62'

flat truss

chord size		overall depth of truss											
tc	bc	16"	18"	20"	24"	28"	30"	32"	36"	42"	48"	60"	72"
2×4	2×4	24'	26'	28'	31'	33'	34'	35'	37'	39'	41'	45	48'
2×6	2×4			28'	31'	34'	35'	36'	38'	41'	43'	49'	53'
2×6	2×6				37'	40'	42'	43'	46'	50'	53'	59'	63'

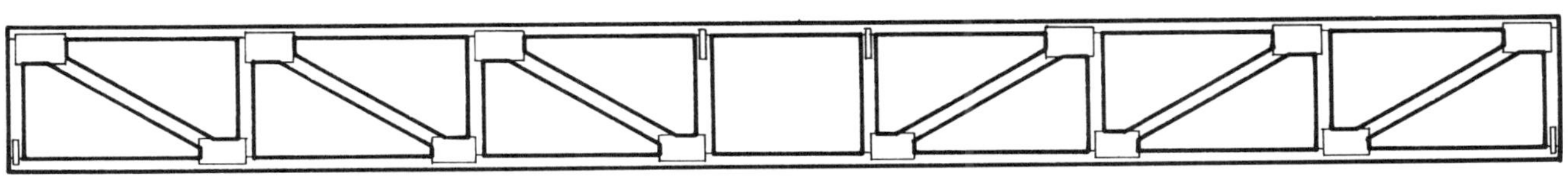

configurations

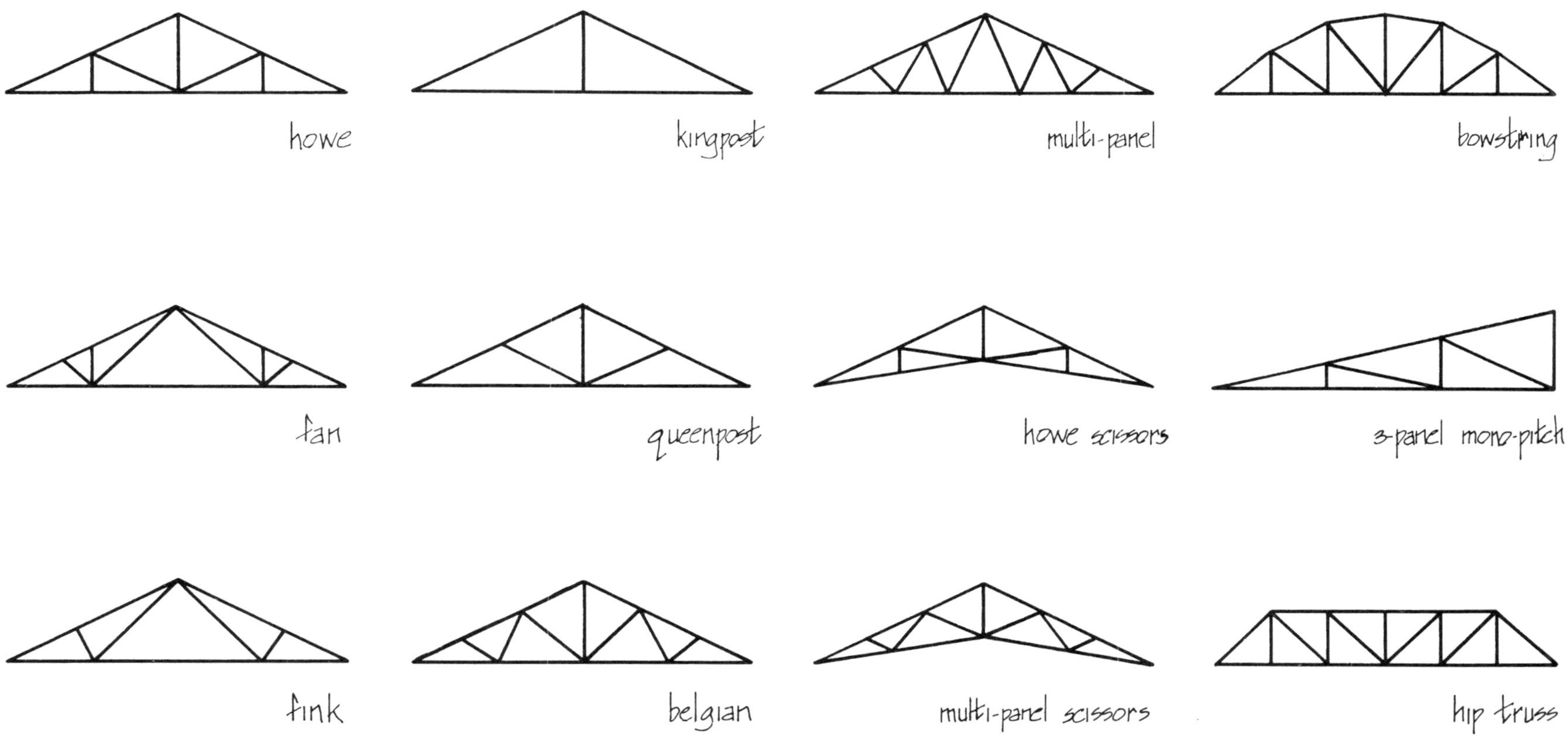

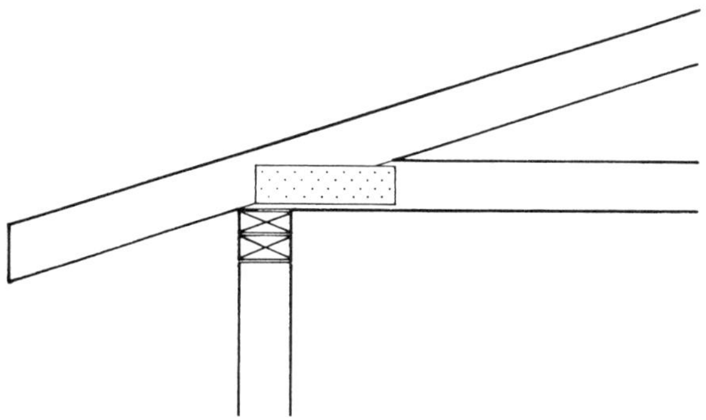

frame wall

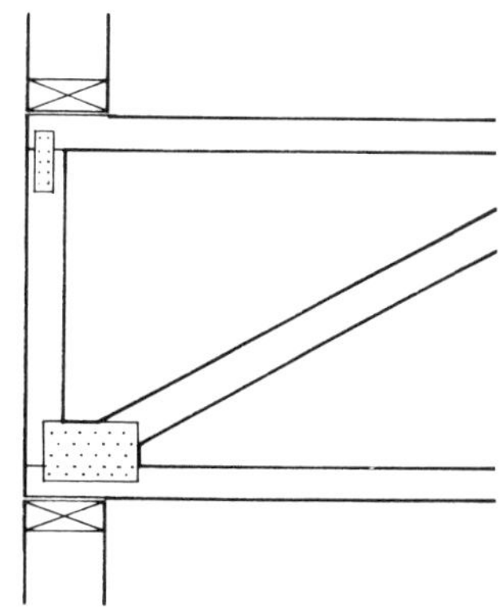

frame wall

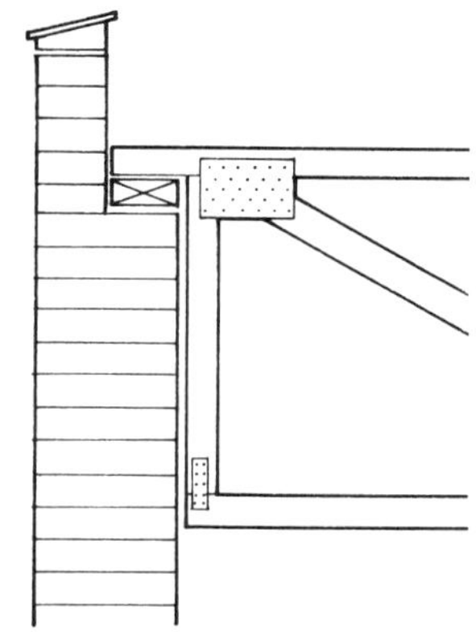

masonry wall

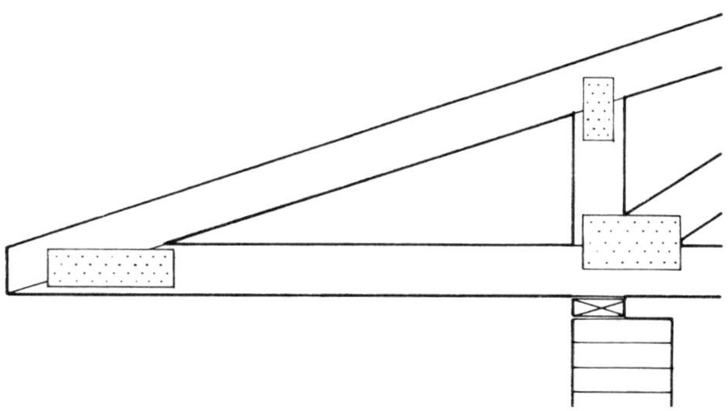

masonry wall

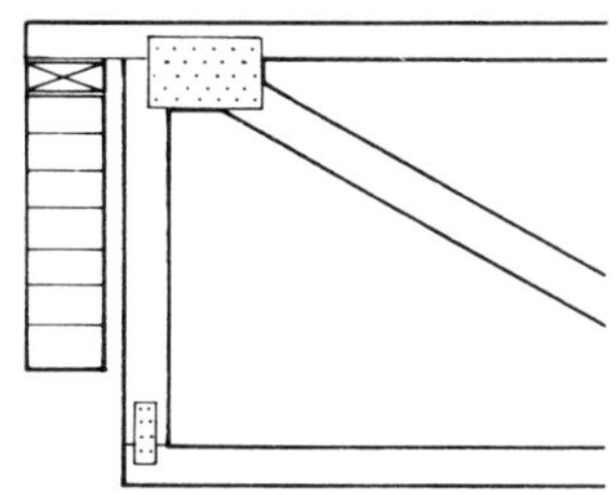

laminated beam

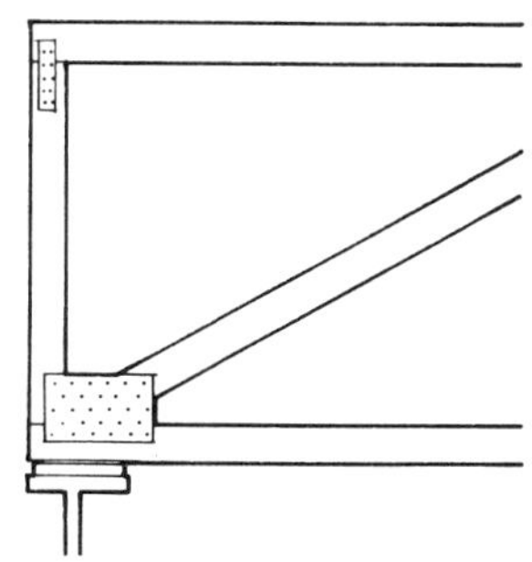

steel beam

METAL SIDING

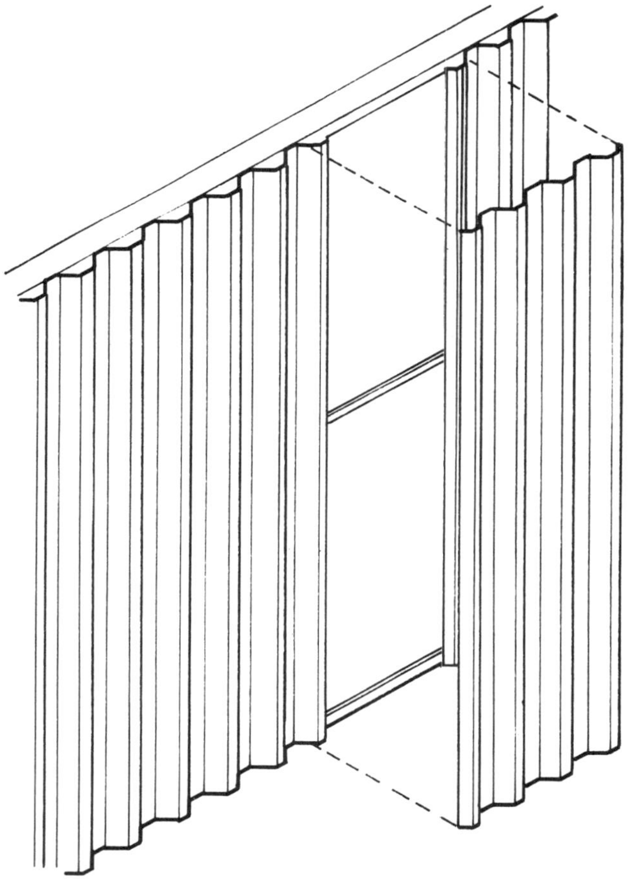

Metal siding, also referred to as metal ribbed panels, is used as non-structural enclosure elements. Formed of sheet steel or aluminum, it provides a lightweight and strong exterior wall surface.

The ribs, which serve a stiffening function, provide a wide variety of profiles with great visual appeal. The metal ribbed panels are available in a range of lengths and widths that allows flexibility in application. Connection of the panels is to secondary framing members or clips attached to primary structure, utilizing corrosion-resistant screws.

Possible finishes include natural aluminum, either plain or textured, baked-on enamel colors, or anodized metal surface. Insulation can be applied to the interior surface either in the field after installation or in the factory. Provisions for doors, windows, or other openings are easily accommodated.

types & sizes

half-hex

	typical sizes
w	41"
d	1¾"
span	14'-0"

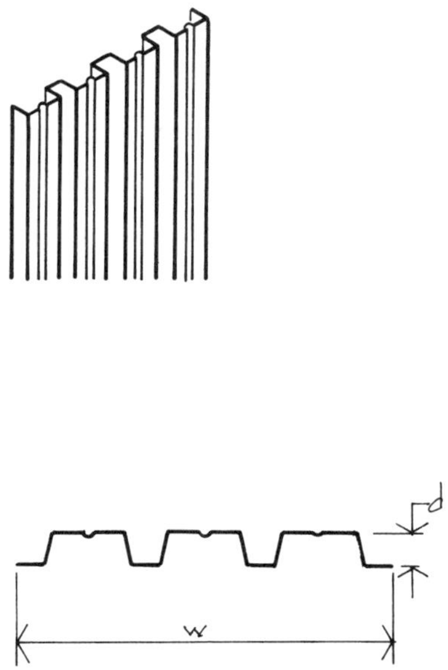

box

	typical sizes
w	24"
d	3"
span	20'-0"

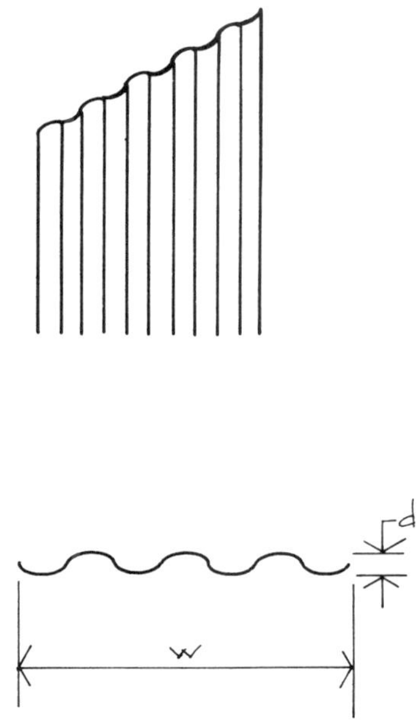

corrugated

	typical sizes
w	33", 47"
d	⅞"
span	7'-6"

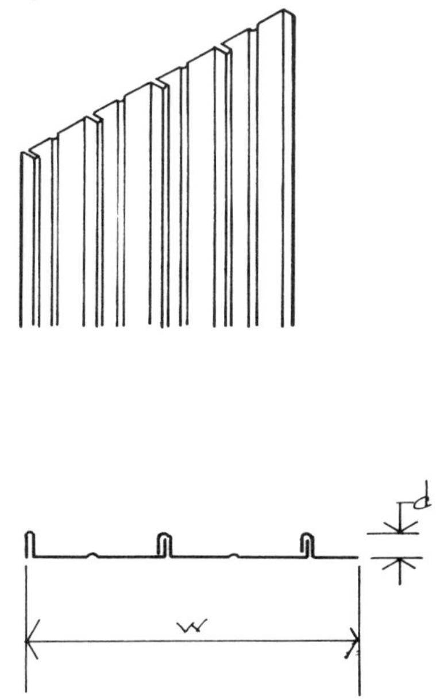

flat with seam

	typical sizes
w	12"
d	¾"
span	10'-0"

profile configurations

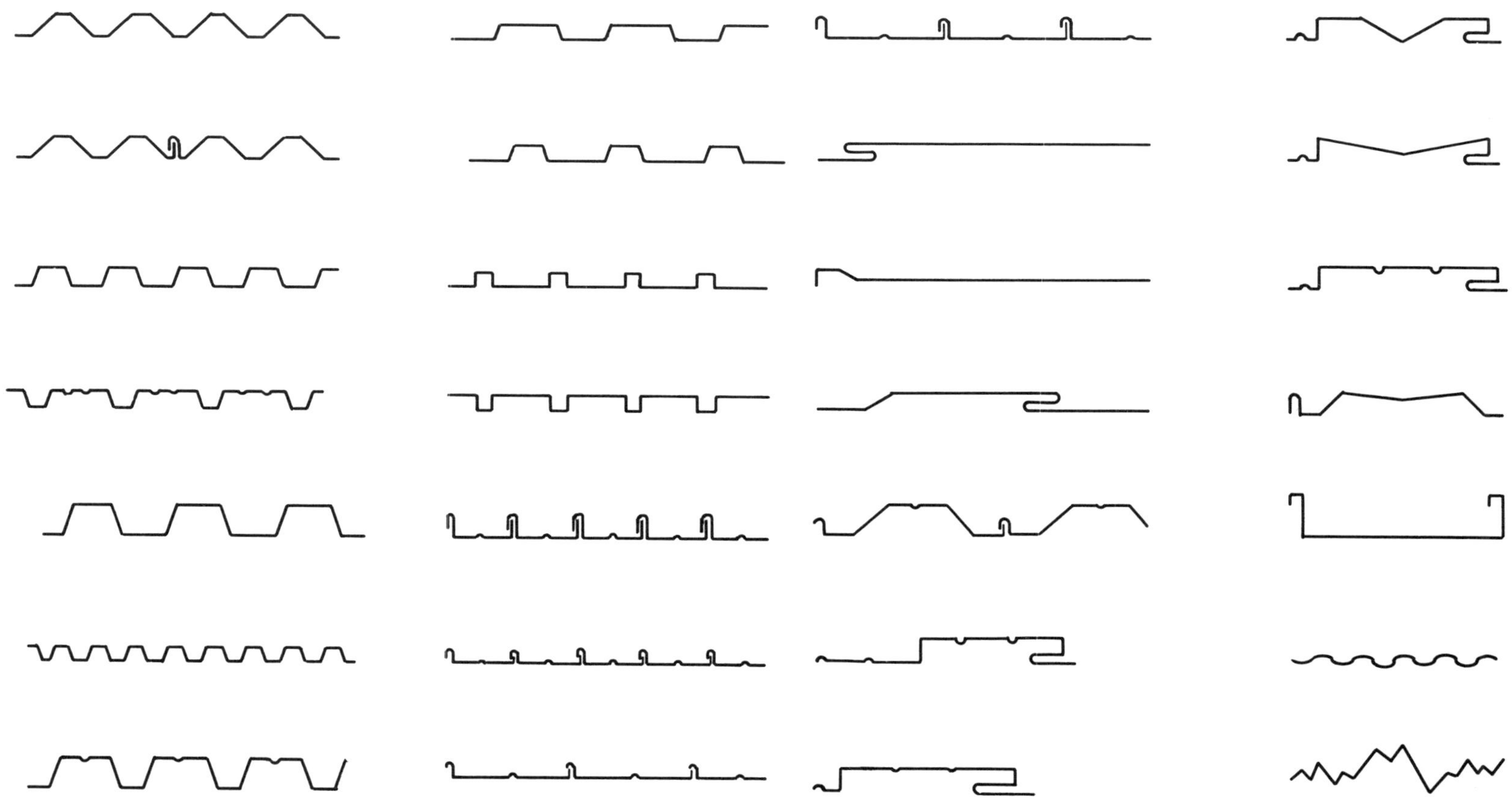

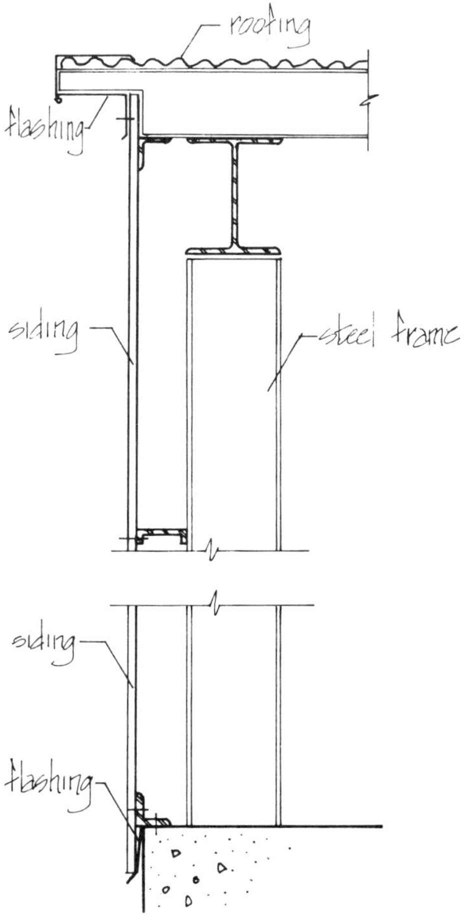
roofing
flashing
siding
steel frame
siding
flashing

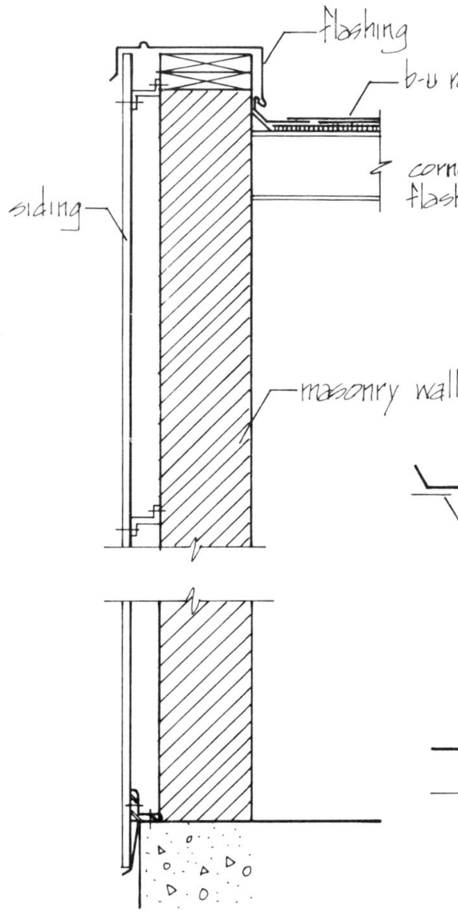
flashing
b-u roof
siding
masonry wall

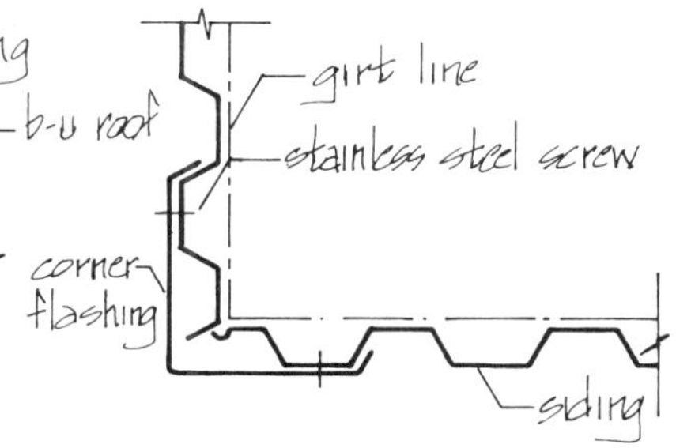
girt line
stainless steel screw
corner flashing
siding

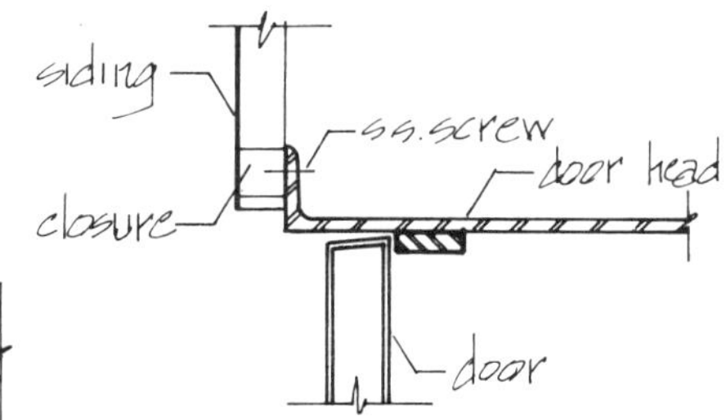
siding
s.s. screw
door head
closure
door

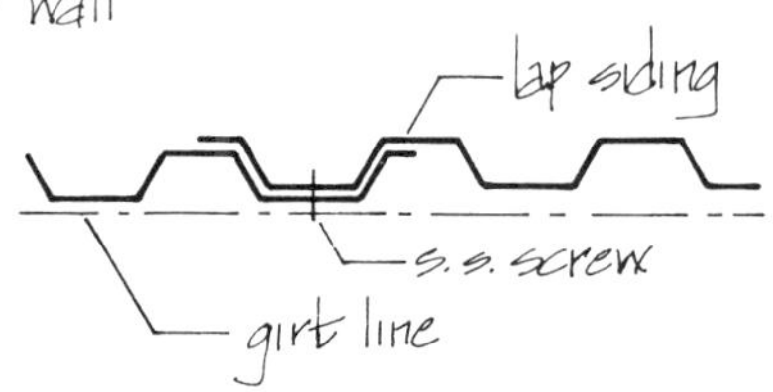
lap siding
s.s. screw
girt line

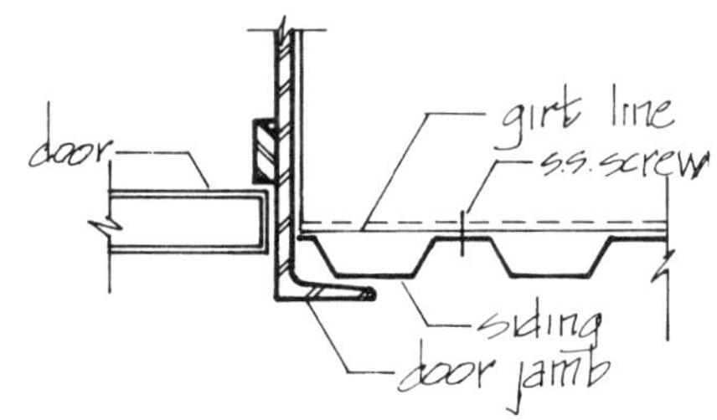
door
girt line
s.s. screw
siding
door jamb

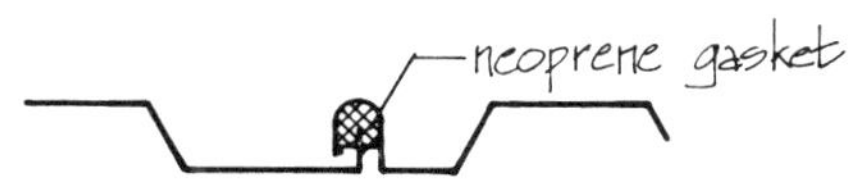
neoprene gasket

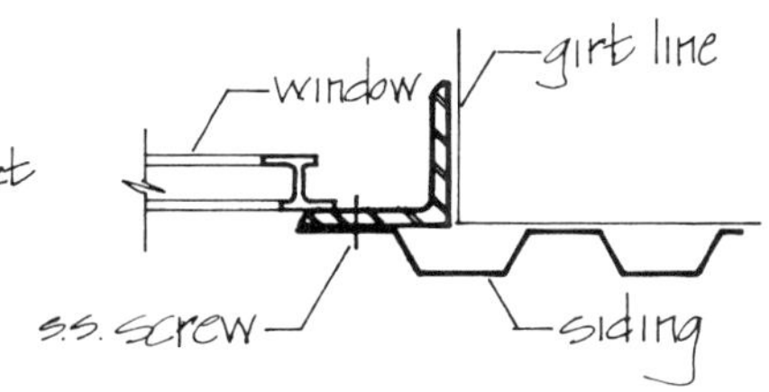
window
girt line
s.s. screw
siding

METAL STAIRS

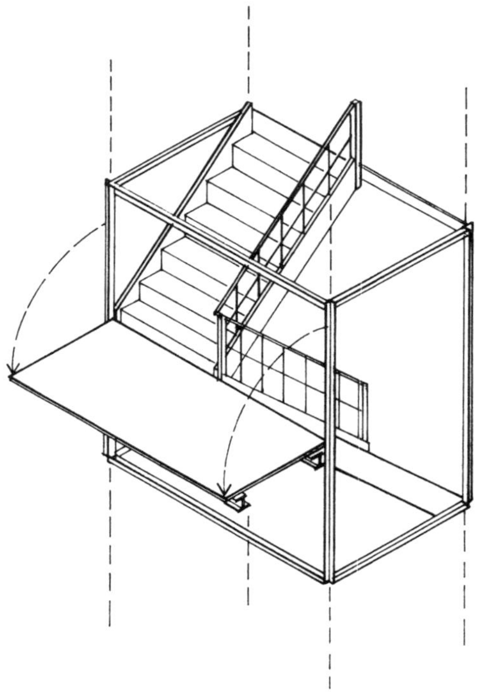

Metal stairs are preengineered, prefabricated, and preerected tower assemblies which provide stairways complete with platforms, center rails, wall rails, and return rails. Towers can be constructed up to four floors or forty foot heights, and towers can be stacked one upon the other to produce stairways of practically any height.

The stair towers are available with standard stairwell openings from 7′-4″ to 8′-0″, and standard floor to floor heights from 8′-4″ to 10′-5″. Special sizes are also available. The platforms, which are commonly made to be hinged down, are constructed with an inverted steel roof deck ready to receive a minimum three inch concrete topping slab. The tower assemblies are commonly of welded construction.

The self-supporting design of the stair units makes them excellent for use as fire stairs. They are delivered finished with one shop coat of rust inhibitive metal primer. Some of the most obvious advantages of prefabricated metal stairs are: ease of installation, economy, time-savings, lack of need for temporary scaffolding, and general availability.

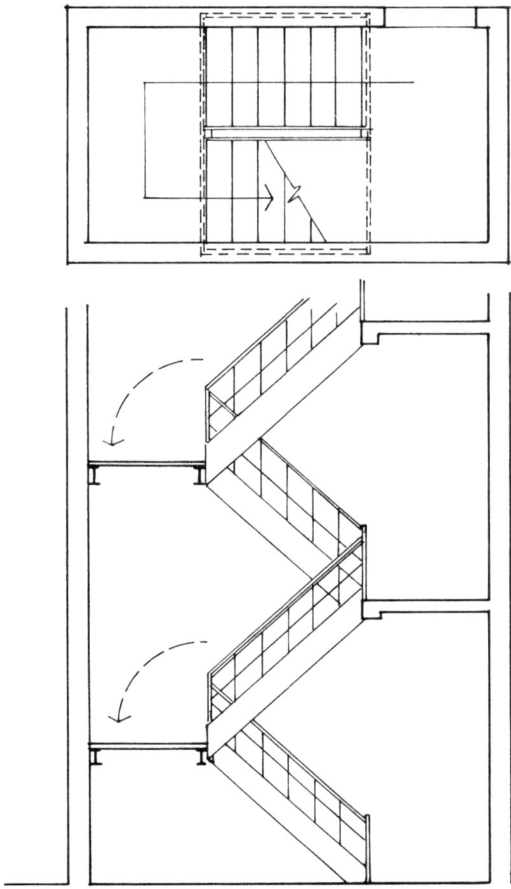

plan

section

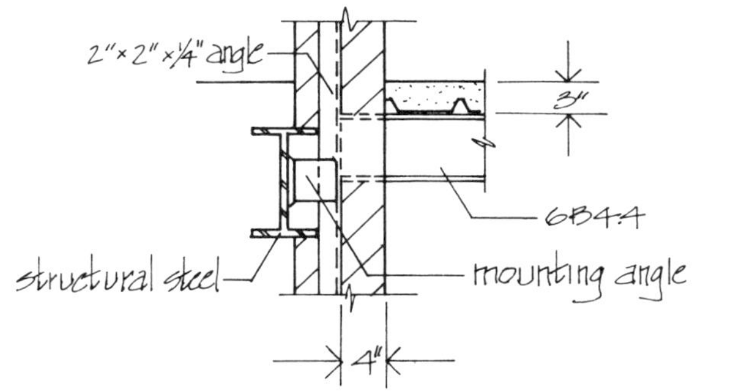

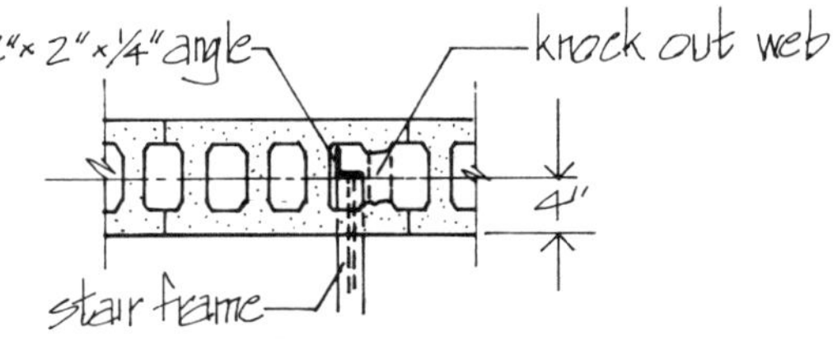

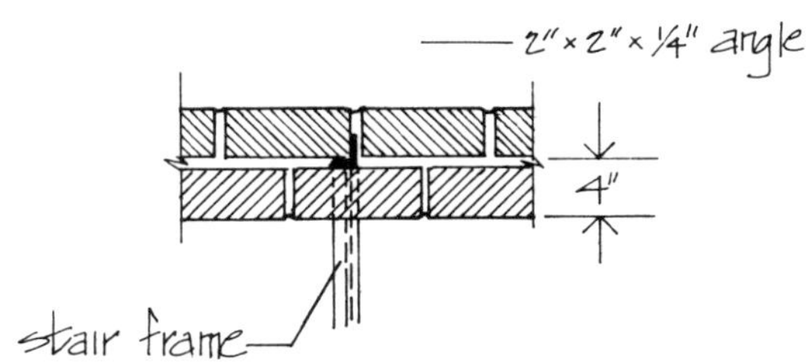

SPIRAL STAIRS

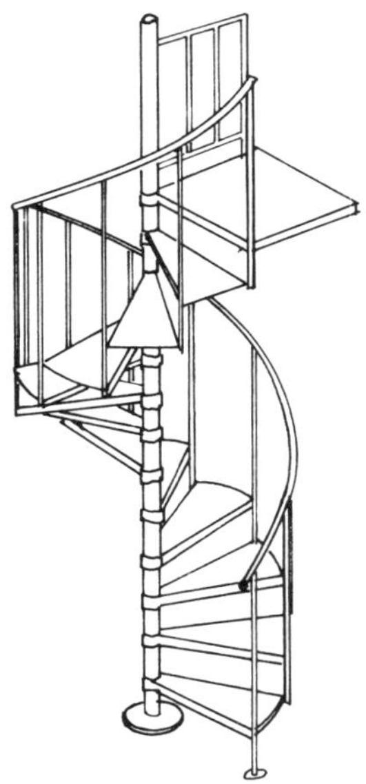

Spiral stairs are prefabricated metal assemblies that provide an inexpensive, economical system of vertical circulation. They can be fully welded and factory assembled, or the component parts can be assembled at the job site.

The stair tread can be 22½° or 30°, providing 16 treads per circle and 12 treads per circle respectively. The treads can have a variety of finishes: plain metal, abrasive coated steel plate, steel pan with lightweight concrete or terrazzo fill, plywood base for carpet, or finish wood tread. The landings can be of quarter circle or square configuration.

The advantages of spiral stairs are based upon the quick and simple method of installation. The center pole is normally fastened to the floor by lag bolts screwed through a base plate, and the platform is secured to the upper floor joists or fascia board by 3/8-inch lag bolts. A variety of sizes and configurations is available.

types & sizes

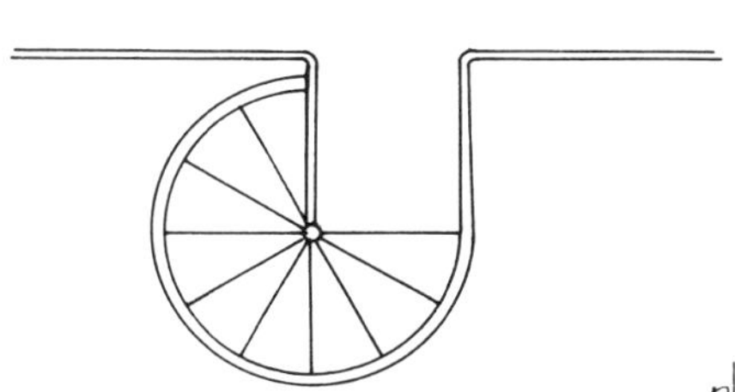

plan

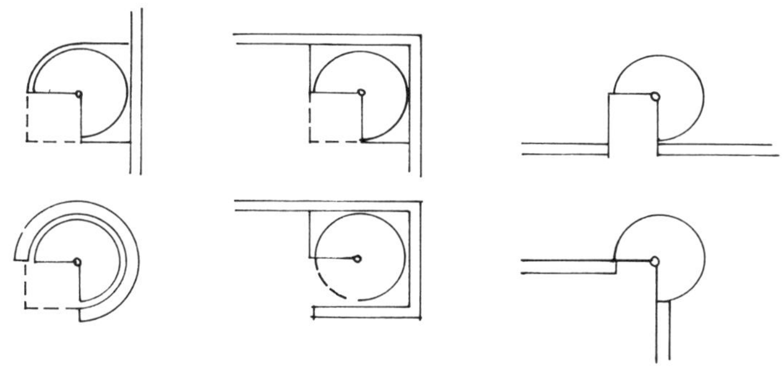

typical stair locations

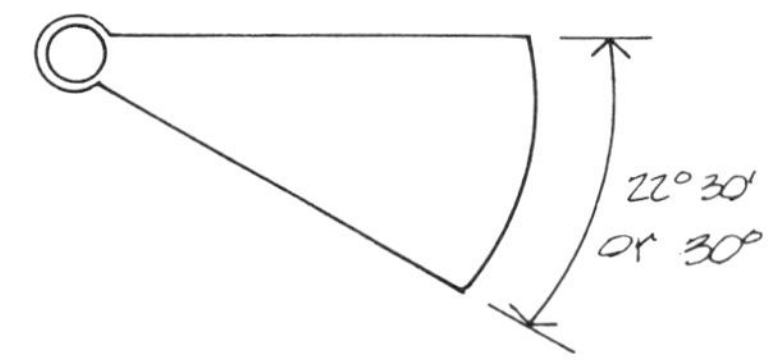

standard tread

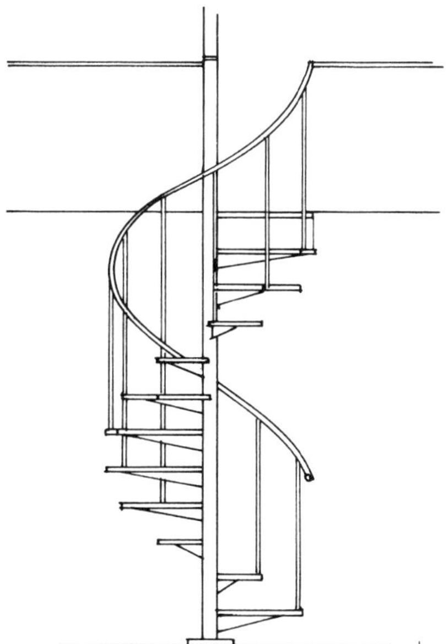

elevation

stair diameter	finished well opening	platform size	width: ℄ of post to rail	pole diameter
3'-6"	3'-8"	22"	17"	3½"
4'-0"	4'-2"	25"	20"	3½"
4'-6"	4'-8"	28"	23"	3½"
5'-0"	5'-2"	31"	26"	4½"
5'-6"	5'-8"	34"	29"	4½"
6'-0"	6'-2"	37"	31½"	4½"
6'-6"	6'-8"	40"	34½"	4½"
7'-0"	7'-2"	43"	37½"	6"
7'-6"	7'-8"	46"	40½"	6"
8'-0"	8'-2"	49"	43½"	6"

tread degree	treads per circle	minimum riser height	headroom
22° 30'	16	7½"	7'-6"
30°	12	8½"	6'-4"

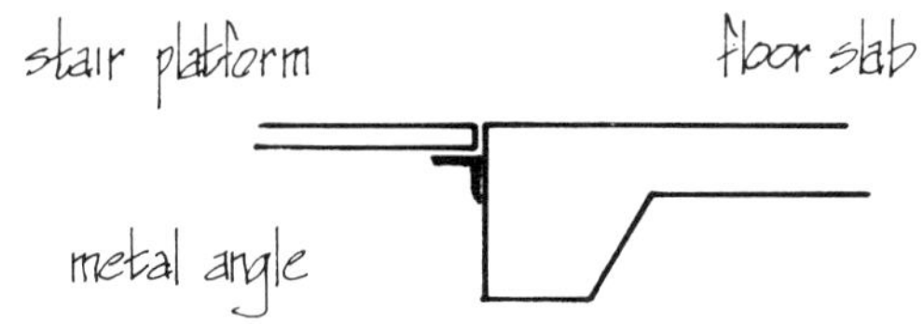

floor connection

CONCRETE STAIRS

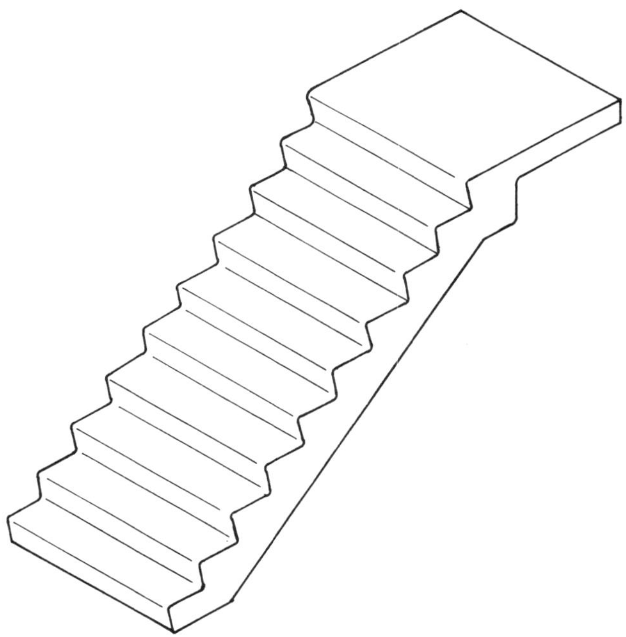

Concrete stairs are precast, reinforced concrete units forming complete flights of stairs. They are preengineered as slabs and require structural analysis of load, support, and span conditions.

The form of the stair, as well as the individual treads and risers, may vary as desired. Various metal nosings are available to protect the edges of the stairs. Treads for concrete stairs may be of metal, rubber, or vinyl. Other finishes are also possible — slate, marble, terrazzo, or carpeting.

The advantages of precast concrete stairs are their fireproof construction and the lack of costly and elaborate formwork required at the site. Their most common uses are in conjunction with other precast building elements.

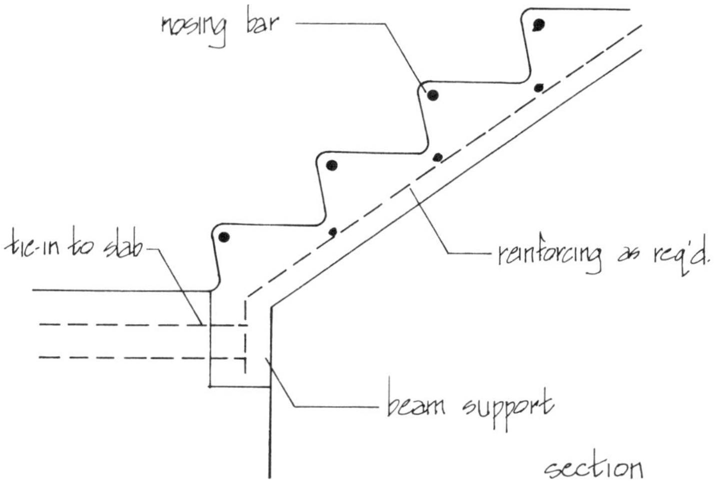
nosing bar
tie-in to slab
reinforcing as req'd.
beam support
section

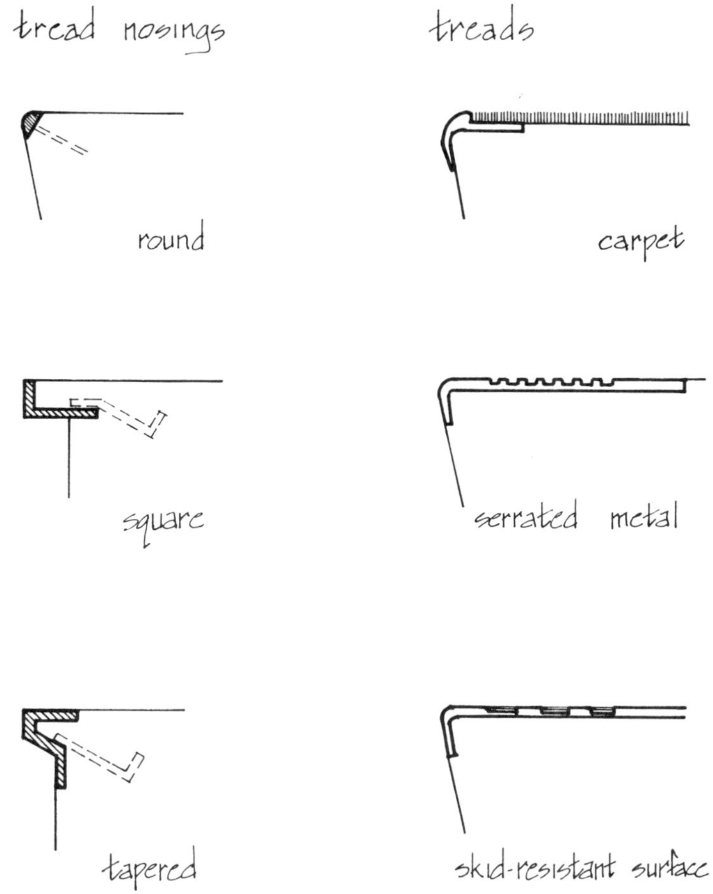
tread nosings
treads
round
carpet
square
serrated metal
tapered
skid-resistant surface

SERVICE MODULES

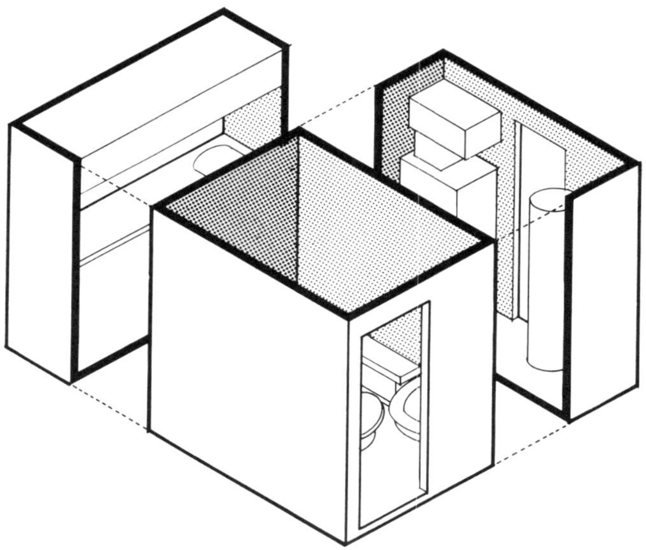

The use of the prefabricated 'wet core' or utility module is intended to maximize efficiency on the building site by consolidating all of the specialized functions of the residential building into a compact, self-contained unit. It is almost exclusively available as a three-dimensional module sized for easy overland transport and erection. Installation of the module is by crane, either lowering or sliding it into place according to its application. The modules are available in finished form including factory installation of all fixtures, appliances and even wall and floor finishes. Allowances are made of necessity for local on-site inspections of all plumbing and electrical equipment.

Standard 'wet' modules include one, one and one-half, two, or two and one-half baths, utility areas, kitchen wet walls, and in the larger modules, enclosed kitchen areas. Options available include a laundry area, mud room, powder room, and storage space. Construction is generally of wood or metal frame and can be load-bearing or non-load-bearing according to the project requirements. Standard products include one- or two-storey modules intended for single-family residential use, as well as two- and three-storey stacking units and non-load-bearing modules for higher density and institutional applications.

plans

variations

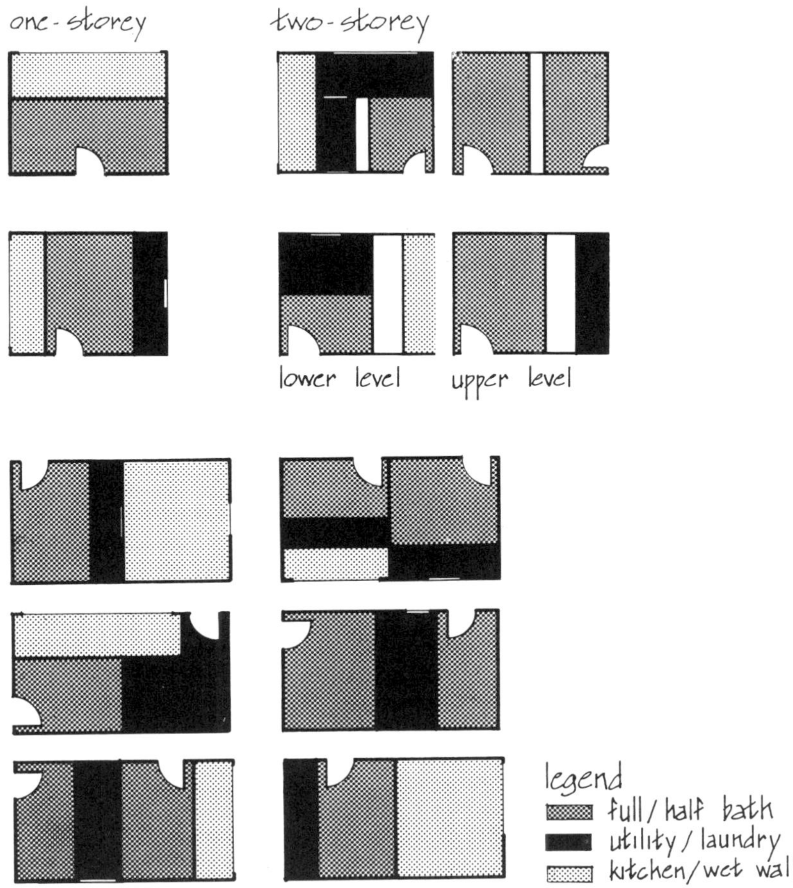

applications

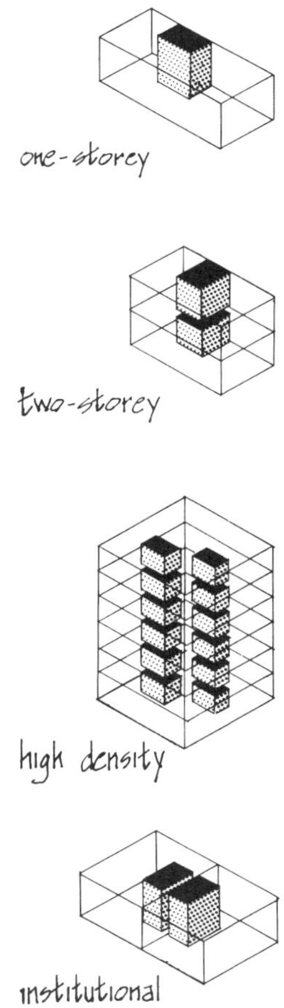

METAL DECKING

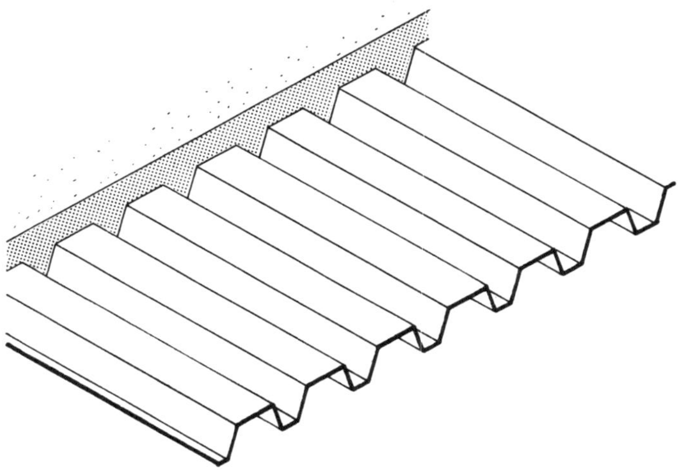

Metal decks for roofs or floors have found a wide application in modern construction. Their use can facilitate the construction process by eliminating the need for separate formwork, since the metal decking serves both as reinforcing structure and formwork for a concrete topping slab. Metal decking is manufactured from roll-formed, structural grade, sheet steel in a variety of gauges. The configurations available are either corrugated or ribbed. They are manufactured in a variety of lengths up to 48 feet. The spans are generally from 5 feet to 20 feet. The specific sizes, spans, gauges, and finishes vary with manufacturer. Possible finishes are galvanized, prime-painted, or uncoated.

The sheets are easily transported, stored, and erected on the job site. Connections to main structural system are usually by welding, although other fastening devices are available.

The ribbed sheets can be used not only as simple floors or roofs, but with the addition of a flat bottom plate, they can be formed into raceways for electrical lines. Also, the ribs can be filled in with glass fibers or other acoustical insulating material to form sound-deadening barriers.

Applications of metal decking include commercial, office, industrial and warehouse buildings, where the decking is combined with a concrete topping slab.

types & sizes

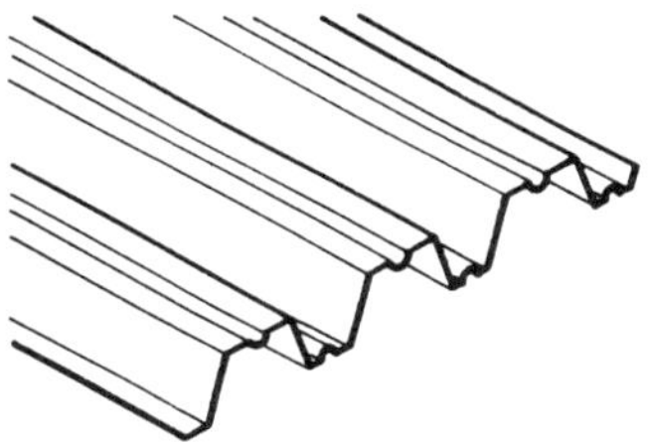

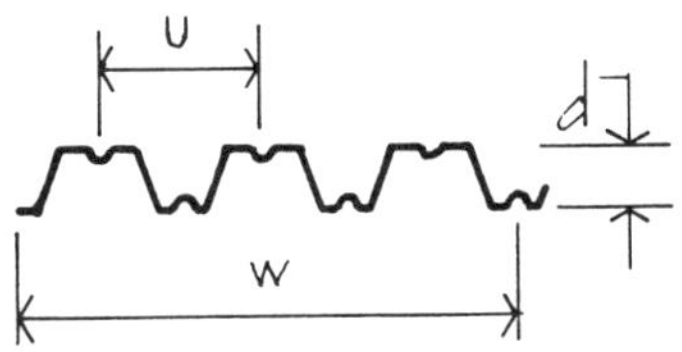

wide-span

w	u	d	span	slab thick.
24"	8"	3"	10'-20'	5½"-7½"
30"	6"	1½"	5'-12'	4"-6"

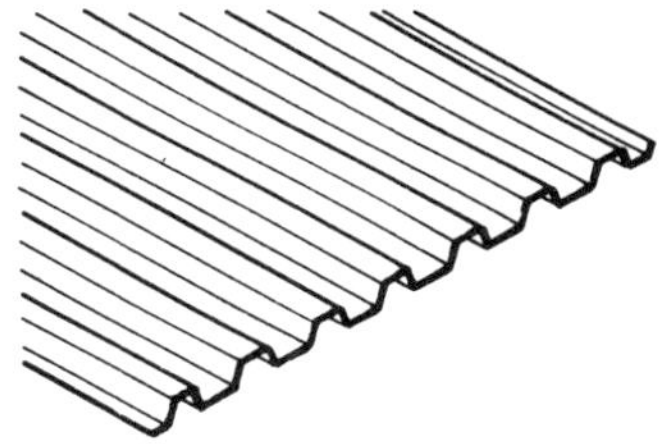

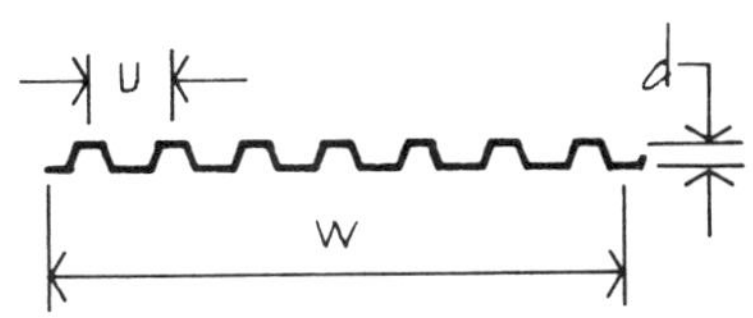

standard

w	u	d	span	slab thick.
30"	2½"	19/32"	2'-5'	2½"-4"
28"	4"	1"	2'-8'	3"-5"
27"	4½"	1⅜"	4'-10'	3½"-5"

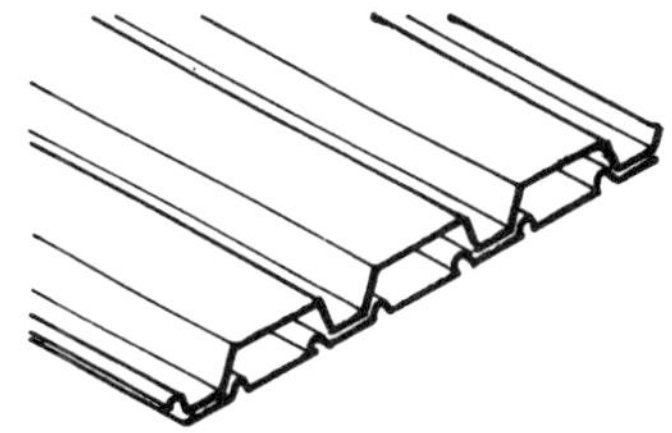

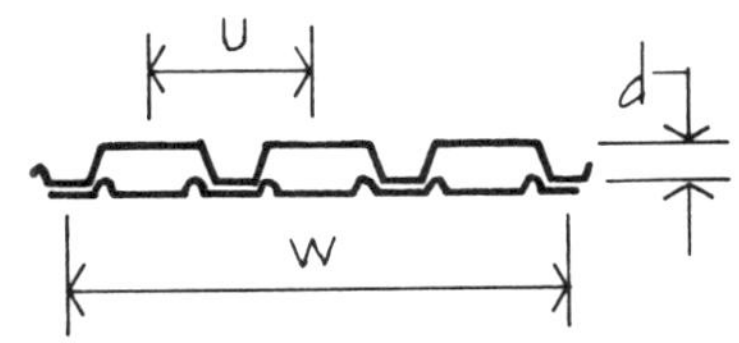

bottom closure plate

w	u	d	span	slab thick.
24"	6"	1½"	7'-13'	4"-6"
30"	8"	2"	8'-14'	4½"-6½"
24"	8"	3"	10'-16'	5½"-7½"

details

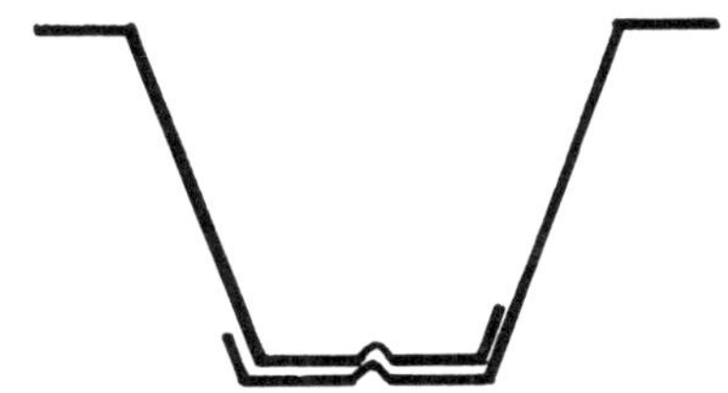

overlap detail

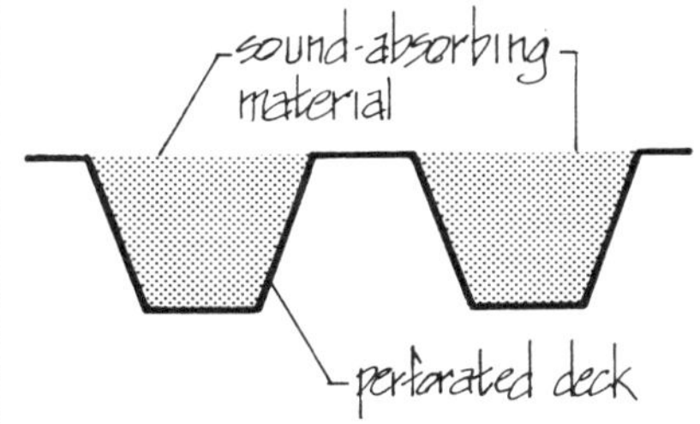

soundproofing

CONC. ROOFING

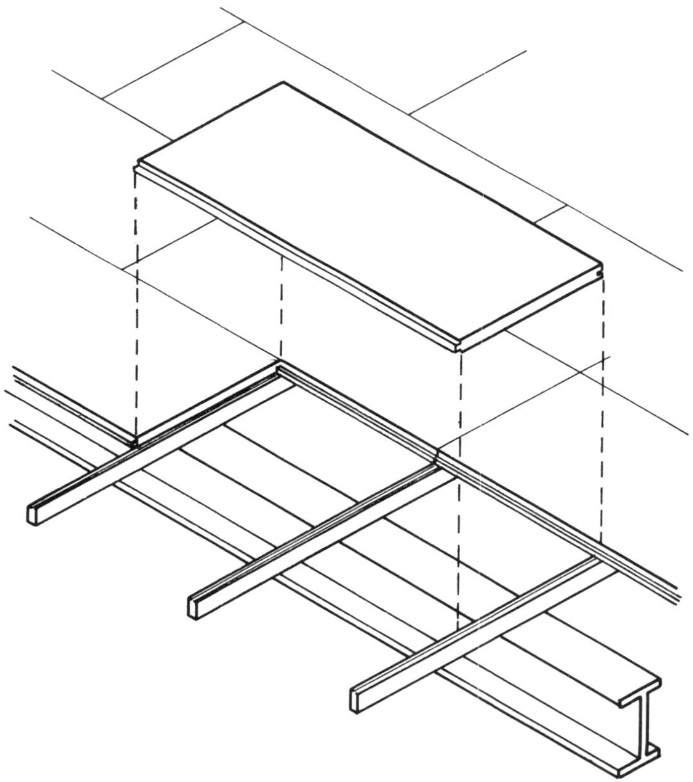

Concrete roof deck is made of structural lightweight concrete in small-sized units of variable configuration. The most common forms are flat slab units, tongue-and-groove planks, rabbeted tiles, and channel slabs. A composite deck, composed of chemically treated mineralized wood fibers and Portland cement, is an alternative.

The sizes and spans of the various types of decking are dependent upon the actual configuration, but are generally in the range of 2 feet widths and 9 feet spans. The roof deck is attached to purlins or joists by a variety of hardware clips, available for steel, concrete, or wood members. Excellent sound absorbing qualities and high insulation values are a result of the porous quality of the concrete and wood fiber composition. Good fire ratings are also a major characteristic.

The deck is easily installed, involving only dry placement and fastening of clips. The joints between the slabs can then be sealed with mastic and felt (if a butt joint) and built-up roofing then applied. The underside of the deck provides a smooth, light-reflecting ceiling which can be left without further treatment. Also, acoustical treatments of different types can be cast in the underside of the deck, or other finishings can be applied after installation.

Concrete roof deck can be used for practically any type of building, from factories to shopping centers.

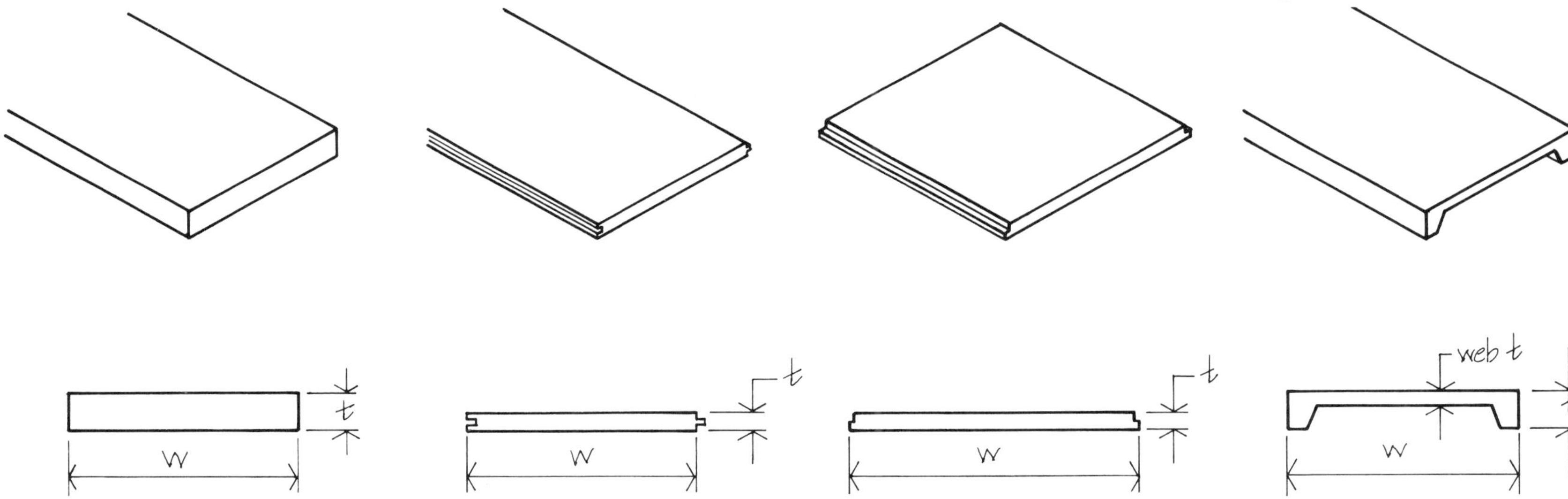

roof deck

w	24"
t	3", 4"
span	6'-0", 8'-4"

roof plank

w	16", 24"
t	1½", 2", 2½", 3", 3½"
span	24" to 54"
long span	48" to 144"

roof tile

w	32"
t	2", 2½", 3, 3½"
span	32"

channel slab

w	24"
t	2¾", 3½", 4"
web t	1", 1½"
span	6'-0", 8'-4", 10'-0"

joint details

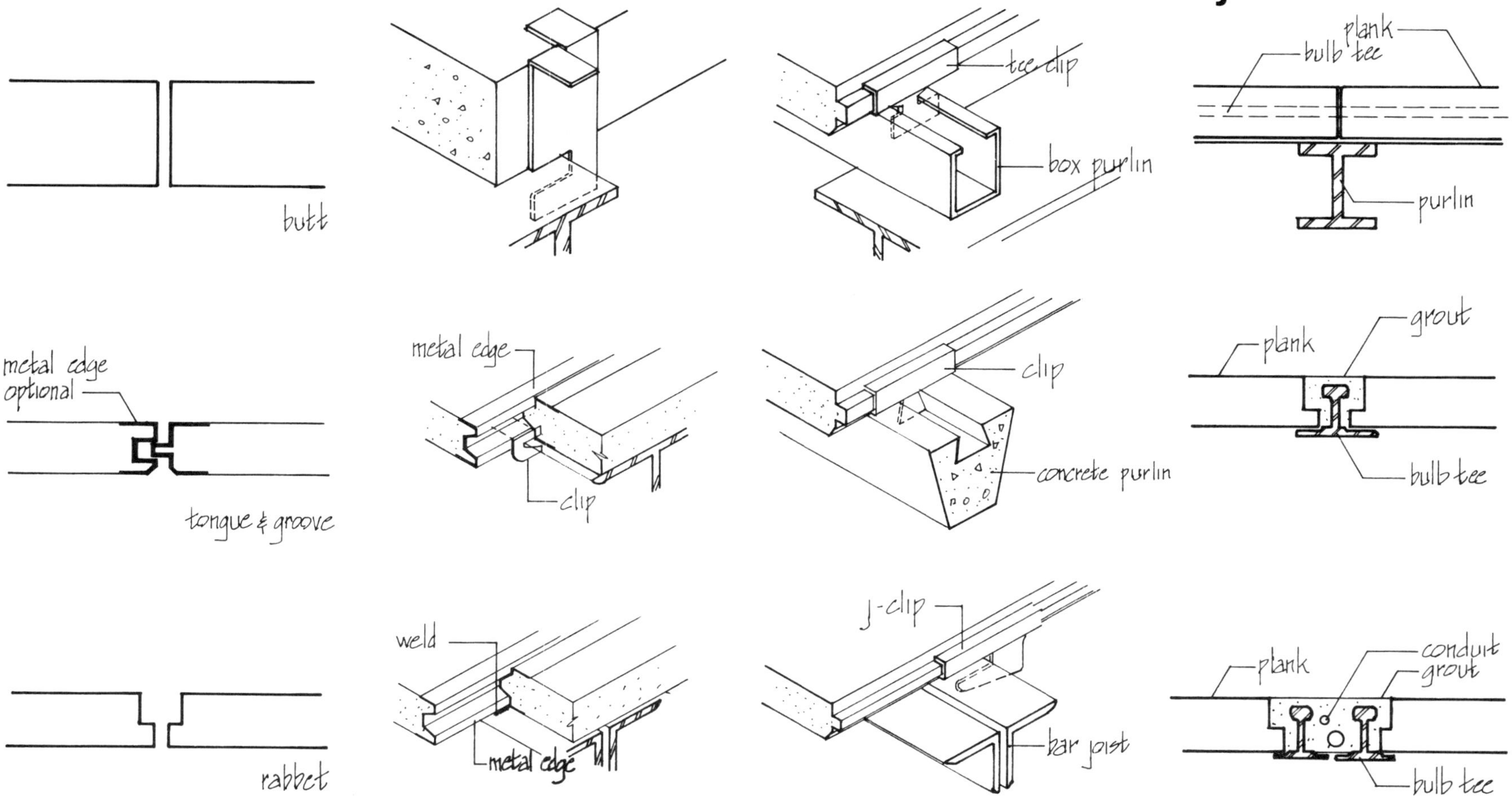

support details

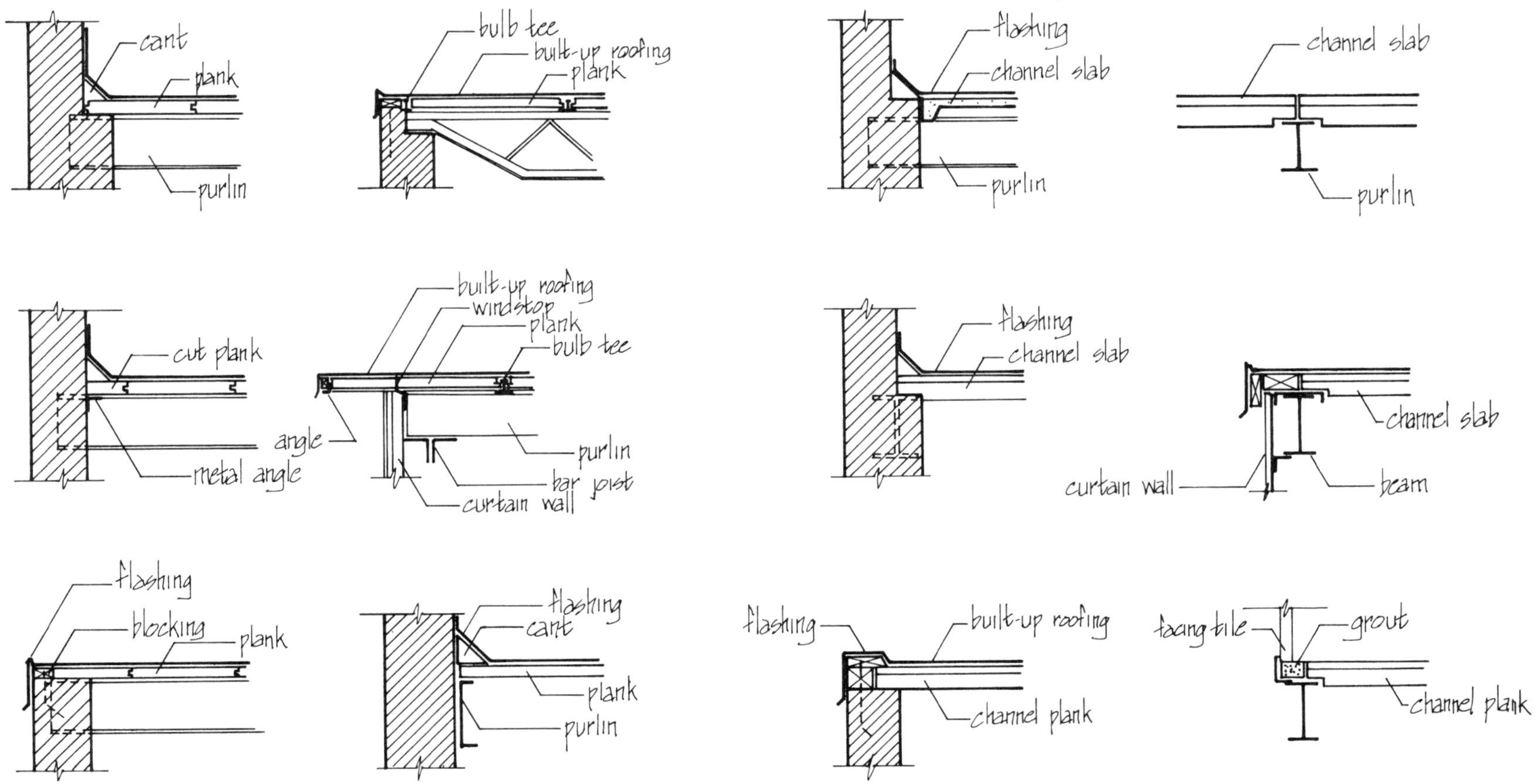

WOOD ROOFING

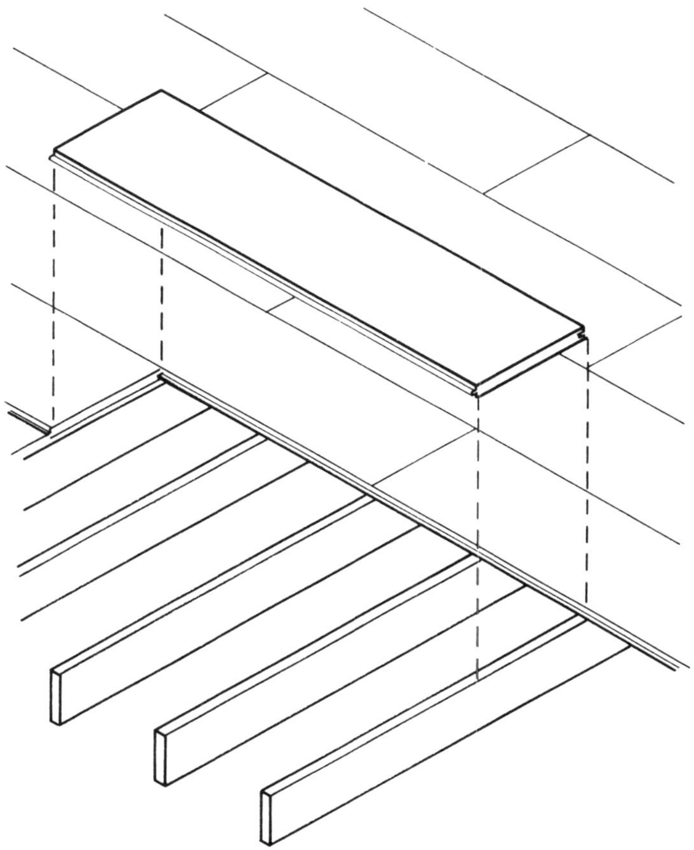

Wood roof deck is made from recycled wood fiber material processed with chemically treated, moisture resistant ingredients into plank units generally 2 feet by 8 feet. They are formed with tongue-and-groove edges, and are usually 15/32 inch in thickness.

The roof deck provides a structural roofing base, thermal insulation, and a finished celing. It can be applied to conventional framing, metal purlins and joists, bulb-T purlins, bar-joists, or prefab trusses. The roof decking can be finished with a wide variety of rigid or asphalt shingles, roll roofing, or bonded built-up roofing. The underside of the planks can be supplied unfinished, prime-painted, or with a vapor barrier film. Practically any type of building can use this form of roof deck.

sizes

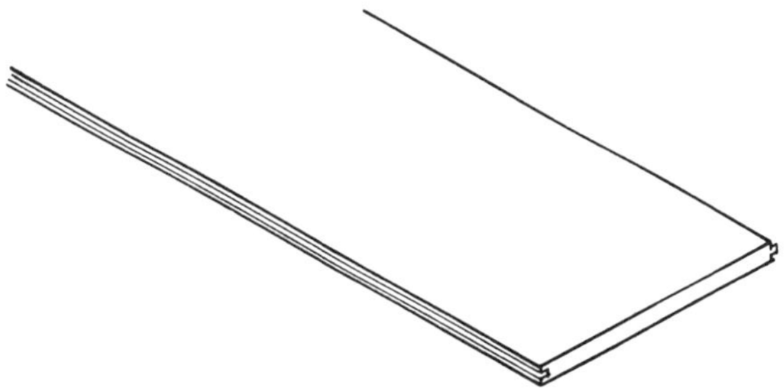

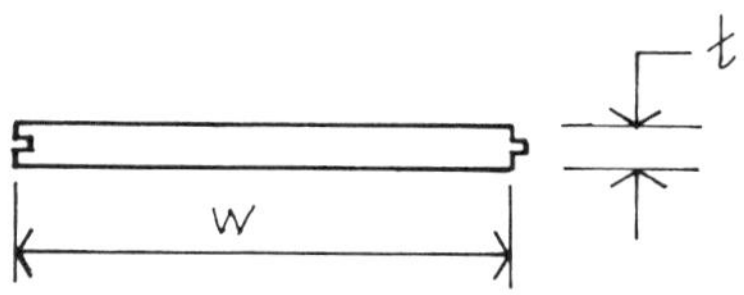

	std. sizes
t	15/16", 1 3/8", 1 3/4", 2 3/8"
w	2'-0"
length	8'-0"

details

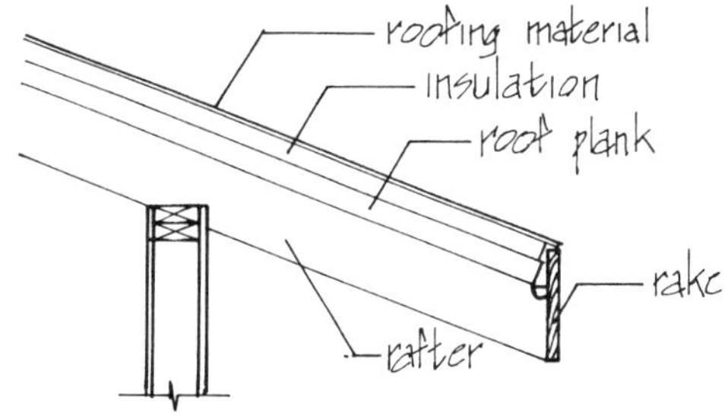

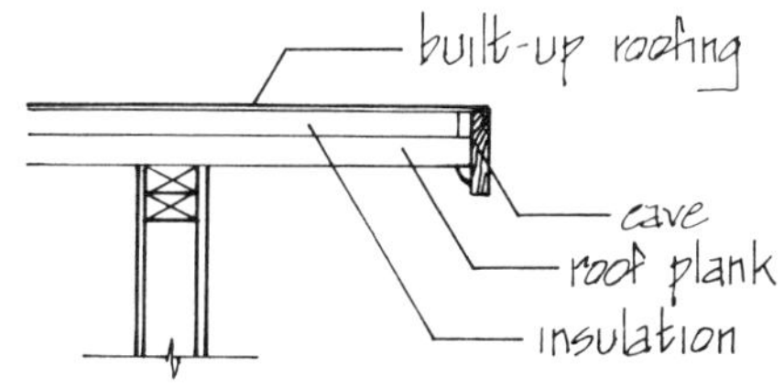

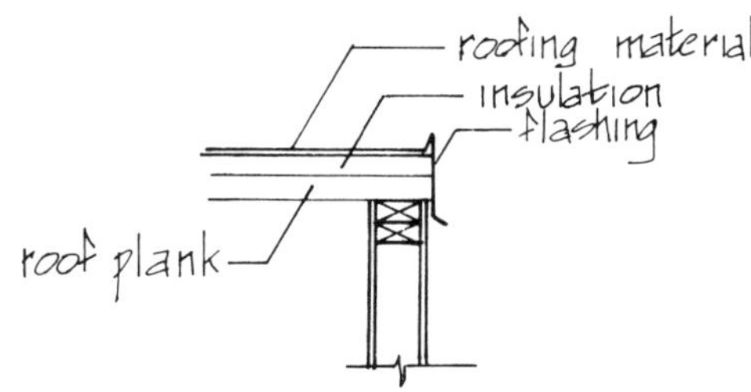

ELEMENTS

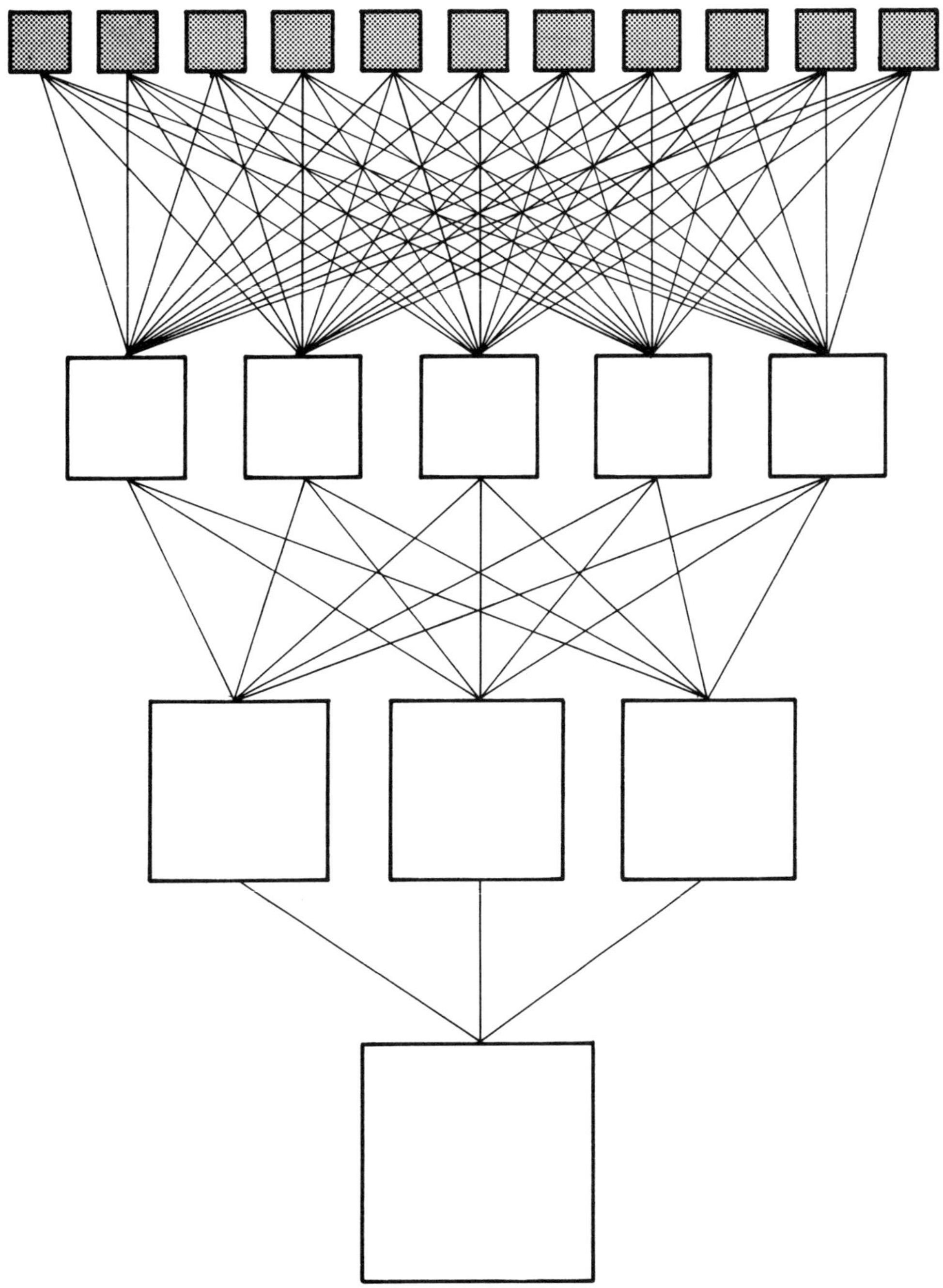

DOORS

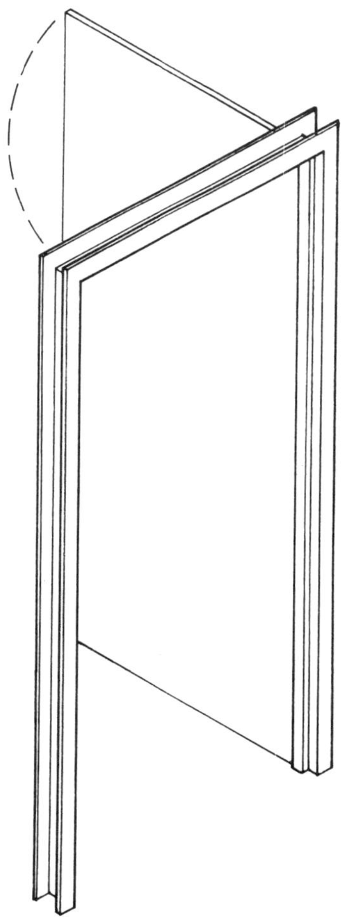

Doors are a common prefabricated building element. They are composed of three basic elements: the door frame, the door itself, and the door hardware. They can be delivered to the site with any combination of these three elements preassembled. Prefabricated doors and frames may be made of wood or metal.

The doors are available in a wide range of styles, patterns, and finishes all prepared at the production facility. Metal doors have a variety of core types, with varying insulation and sound-deadening properties. Wood doors can be solid or hollow core. Special doors such as fire-rated doors, acoustical doors, or lead-lined doors are some of the options available. With the standardization of sizes produced, rough openings can be prepared for the quick installation of the finished door and frame.

types

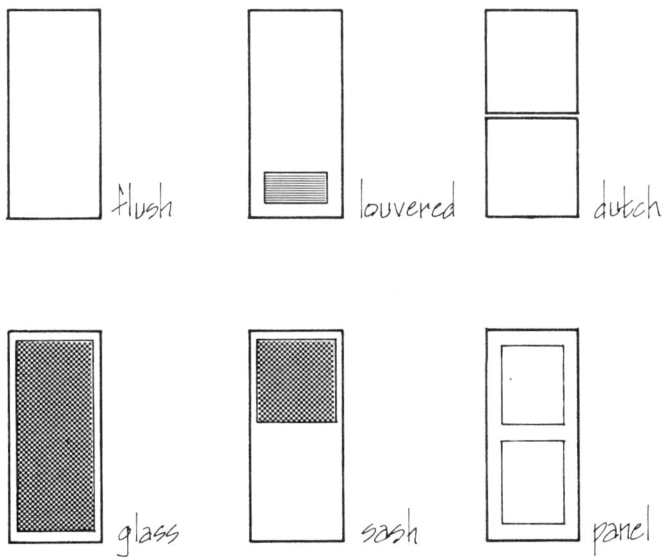

wood door finishes
- hardwood veneer
- hardboard face
- laminated veneer face

metal door finishes
- baked enamel
- vinyl clad
- primed and/or galvanized
- stainless steel or aluminum

sizes

wood doors

width	interior solid core t=1⅜", 1¾" 6'-6"	6'-8"	7'-0"	interior hollow core t=1⅜", 1¾" 6'-6"	6'-8"	7'-0"	exterior solid core t=1¾", 2¼" 6'-6"	6'-8"	7'-0"	exterior hollow core t=1¾" 6'-6"	6'-8"	7'-0"
1'-6"		●			●	●						
1'-8"					●	●						
2'-0"		●	●	●	●	●						
2'-2"					●	●						
2'-4"		●	●		●	●		●	●			
2'-6"		●	●		●	●		●	●			
2'-8"		●	●		●	●		●	●		●	●
3'-0"		●	●		●	●		●	●		●	●
3'-6"								●	●			

metal doors

width	height (t=1⅜", 1¾") 6'-8"	7'-0"	7'-2"	8'-0"
2'-0"	●	●		
2'-4"	●	●	●	
2'-6"	●	●	●	
2'-8"	●	●	●	
3'-0"	●	●	●	●
3'-4"	●	●		
3'-6"	●	●		
3'-8"	●	●	●	
4'-0"	●	●		●

core types

wood

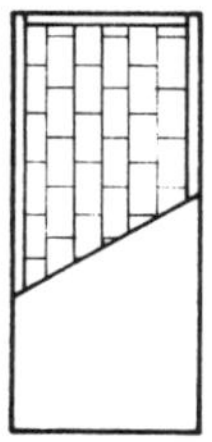

solid wood block
- glued block
- framed block, glued
- framed block, non-glued
- stile & rail

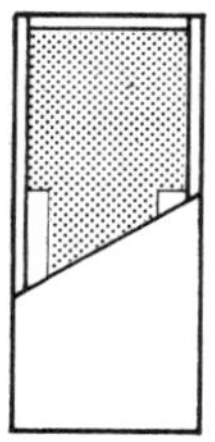

solid composition board
- solid composition board
- mat wood particleboard
- other particleboard
- mineral core

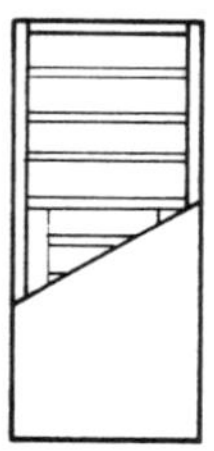

hollow core
- ladder core
- mesh or cellular
- implanted

metal

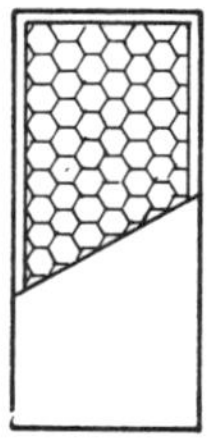

honeycomb

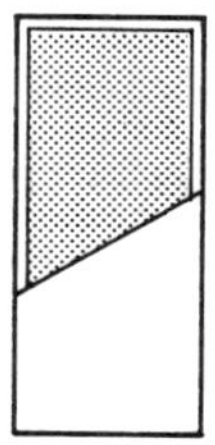

composite
- foam core
  - urethane
  - polystyrene
- mineral fiber

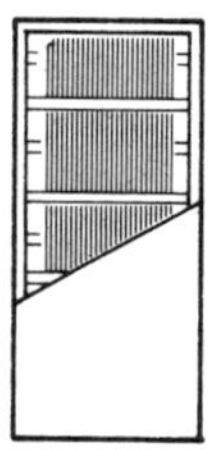

stile & rail
- glass-fiber sound-deadening infill material

frames

wood

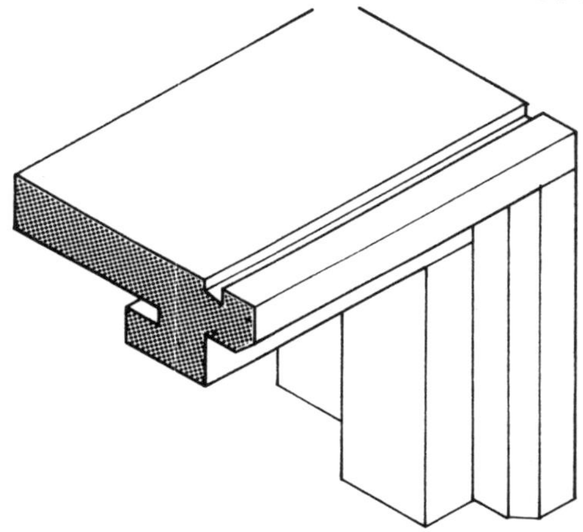

metal

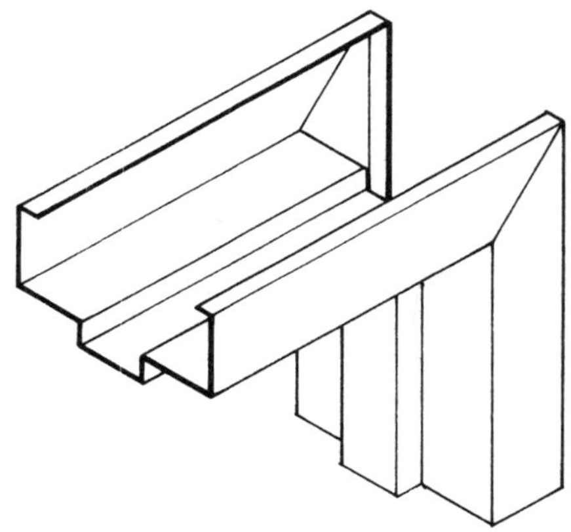

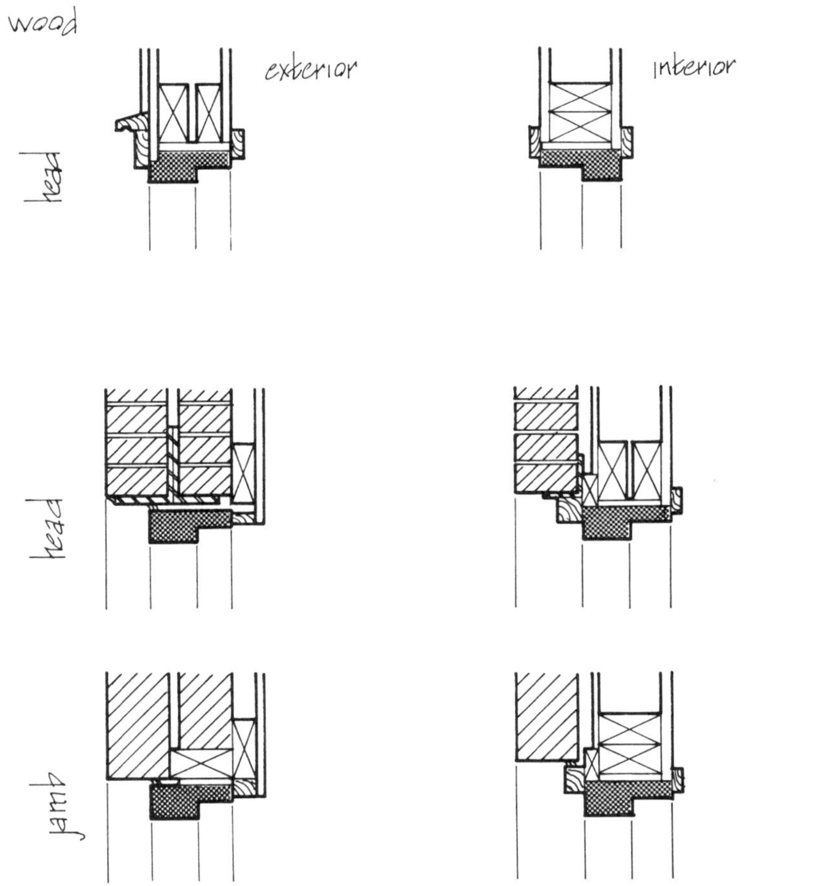
wood
exterior
interior
head
head
jamb

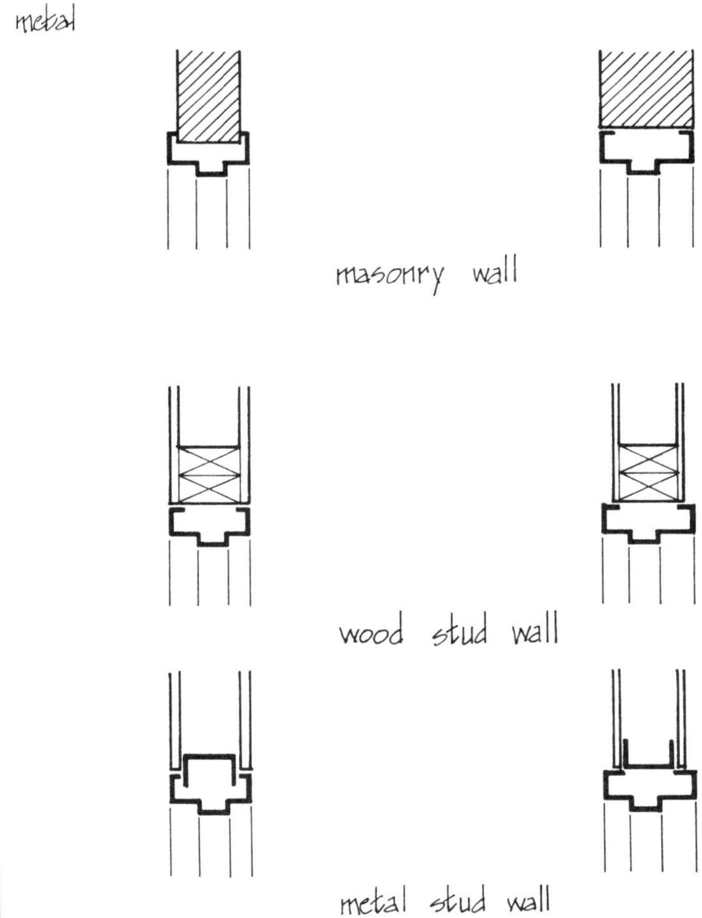
metal
masonry wall
wood stud wall
metal stud wall

WINDOWS

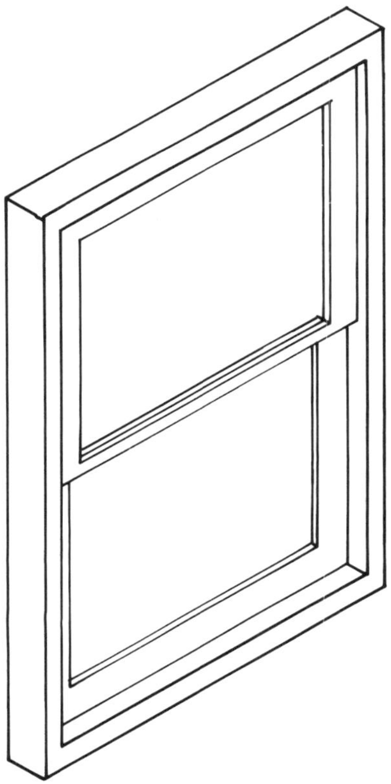

Windows are now almost universally a prefabricated building element. Whether constructed of wood or metal, they arrive at a job site preassembled and ready to install in the rough wall opening.

Stock window units come in a wide range of styles and sizes that can be used for nearly all applications. Wood windows are made of kiln-dried lumber, factory treated with a water repellant preservative. They can be finished with a prime coat, factory painted, or clad with a vinyl or acrylic-coated aluminum. Metal windows can be of aluminum, steel, stainless steel, or bronze.

Fixed glazing consists of a pane of glass, some sort of seal, and the frame or opening. Panes of glass are available in a variety of sizes, thicknesses, and types. The seal is normally a neoprene gasket. The critical factor with fixed glazing is that a smooth contact surface is required, and that the frame or opening must have exacting dimensional tolerances for a true alignment of the glass plane to avoid any possibility of warping.

basic types

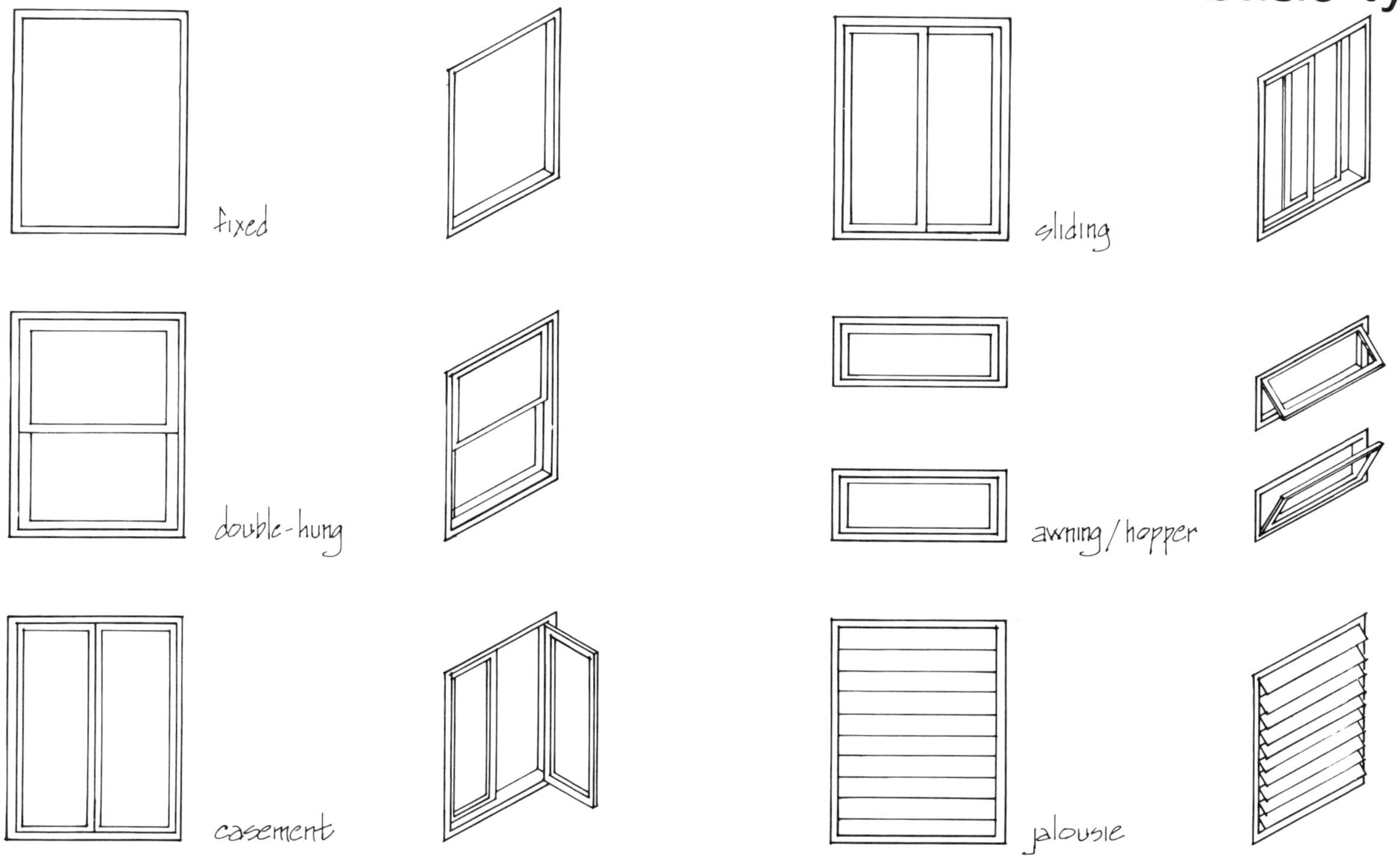

fixed
sliding
double-hung
awning / hopper
casement
jalousie

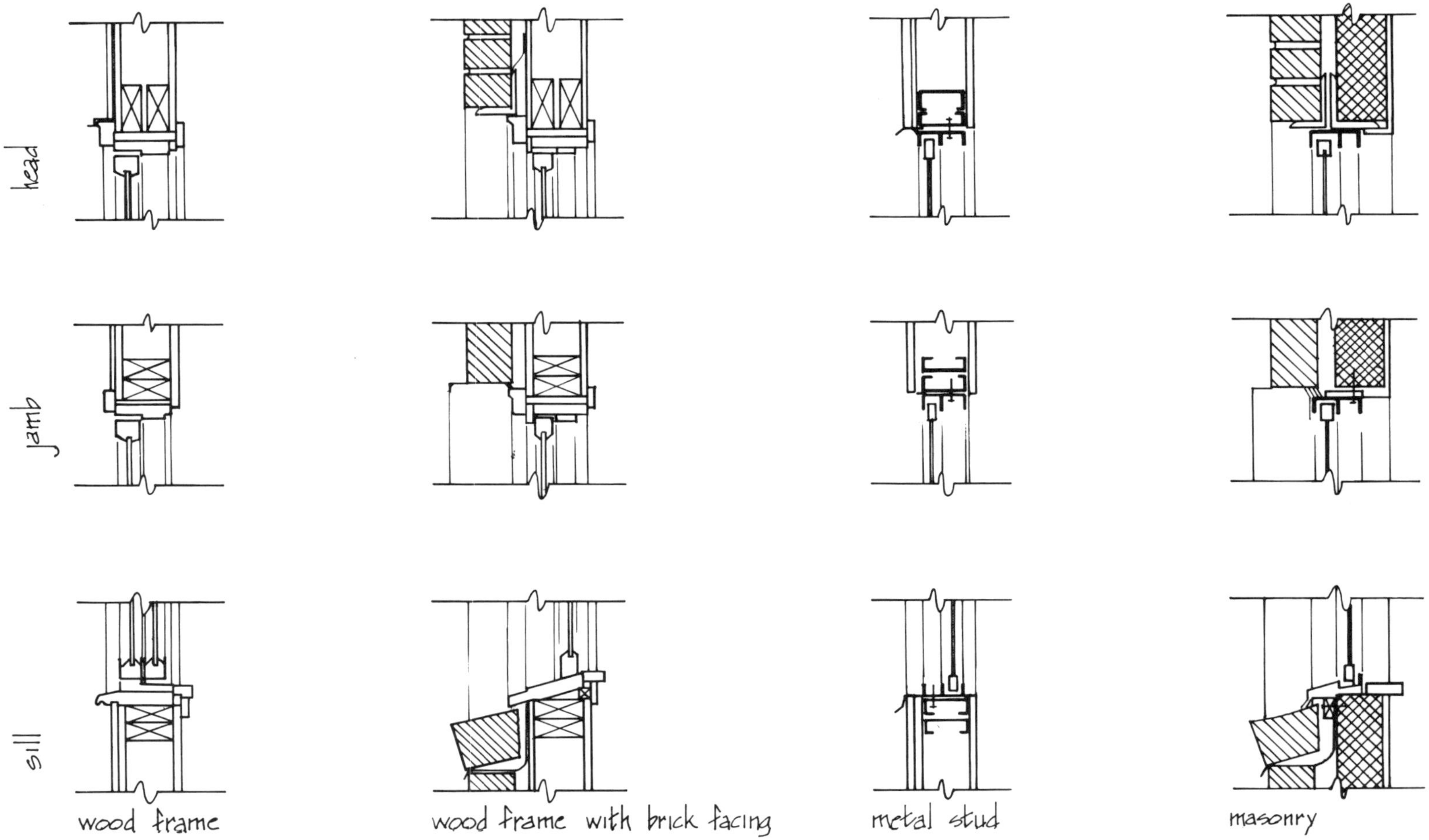
head
jamb
sill
wood frame
wood frame with brick facing
metal stud
masonry

sizes

double-hung

width:	2'-0" to 6'-1"
height:	3'-6" to 6'-10"

casement

width:	2'-0" to 6'-4"
height:	2'-11" to 6'-3"

awning

width:	3'-0" to 4'-8"
height:	1'-11" to 6'-3"

glazing

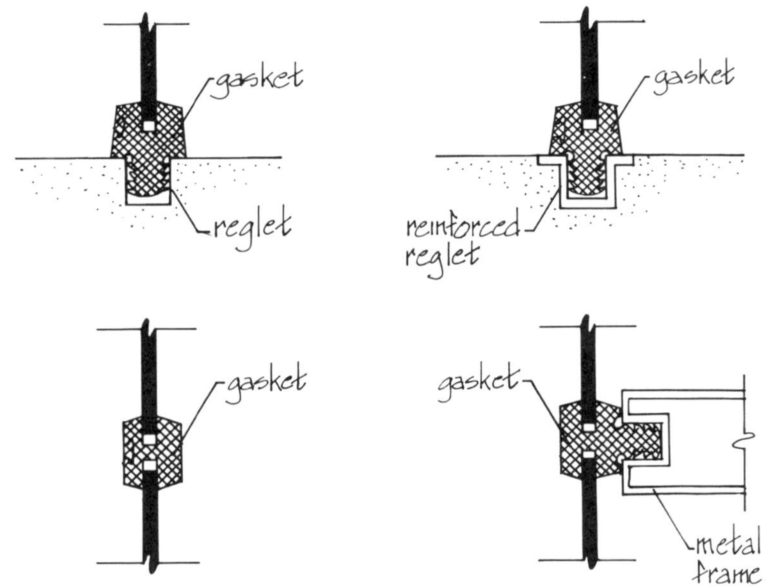

glazing

sheet glass:	1/16" to 7/16"
float glass:	1/8", 3/16", 1/4"
plate glass:	1/8" to 1"

SKYLIGHTS

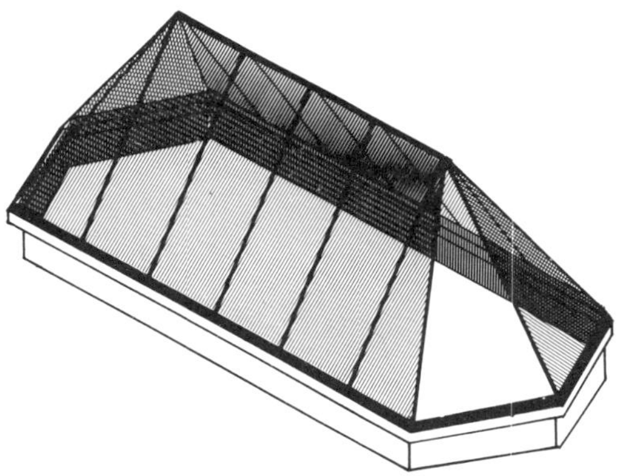

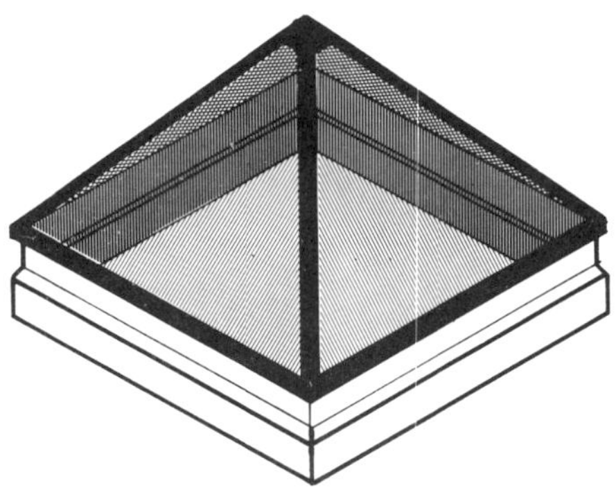

Skylights are prefabricated glazing systems that allow light to enter a room from above. They are composed of a frame, usually made of extruded or tubular metal, and the glazing panel, either glass or plastic. A variety of stock configurations is available, such as domes, vaults, pyramids, and single ridge, and special designs are also possible.

The glazing panels can be clear, tinted, or translucent. Glass panels can be plain, tempered, or wire glass, and plastic panels can be single sheet or sandwich acrylic panels. Skylight frames are usually of aluminum, galvanized or stainless steel, Monel, copper, or lead-coated steel or copper. A curb is required to prevent rain or snow penetration. It may be an integral part of the prefabricated element or may be built at the site. In all designs, whether standard or special, provision must be made for drainage of condensation. This is usually done by having a special lip on the frame member. The connection between the frame and the glazing panel is by means of a neoprene gasket which allows for differential movement and forms a weathertight seal.

The use of skylights can have a profound effect on whatever type of building to which it is applied. The unique quality of lighting it provides is a major advantage, although heat insulation may be a problem.

configurations

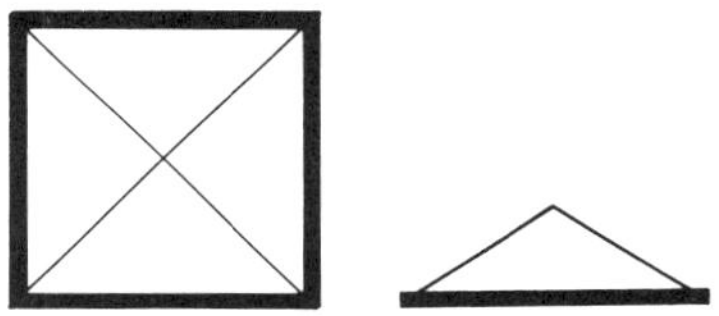

pyramidal

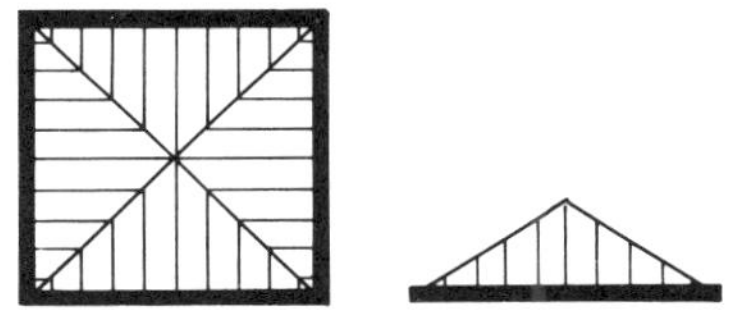

pyramidal

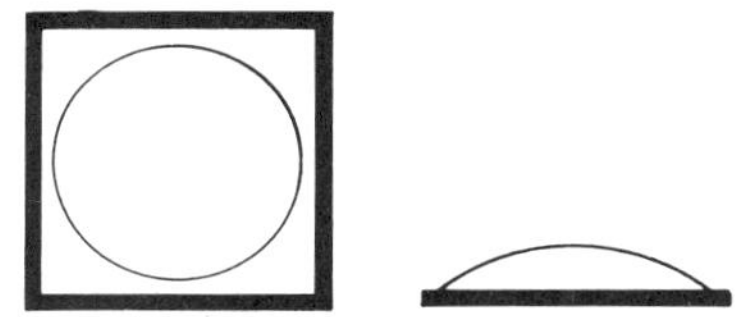

dome

triangular

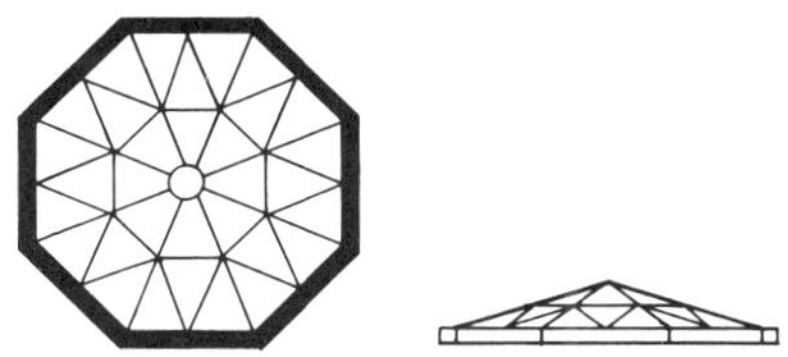

octagonal

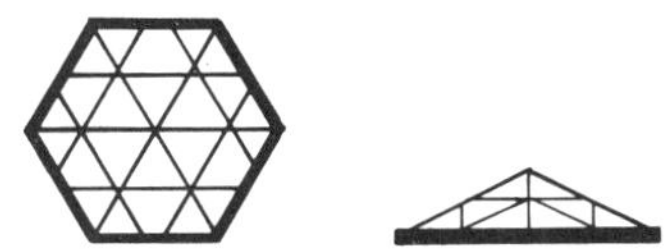

hexagonal

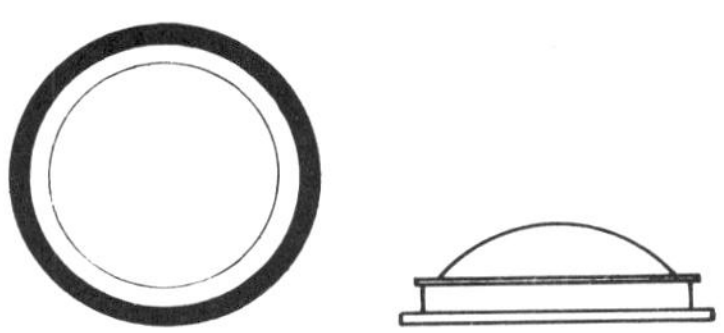

dome

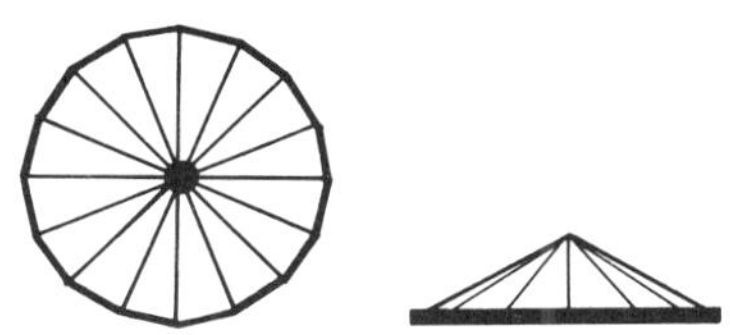

polygonal

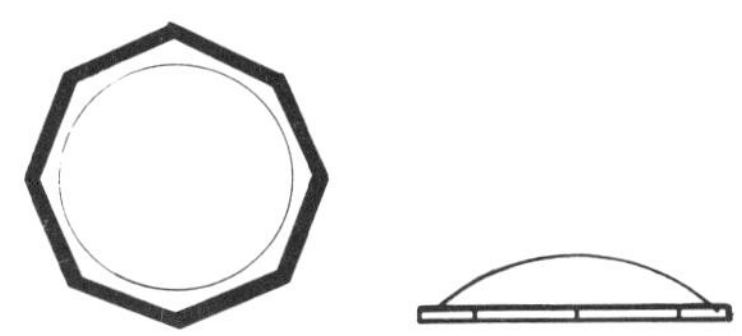

octagonal dome

FIREPLACES

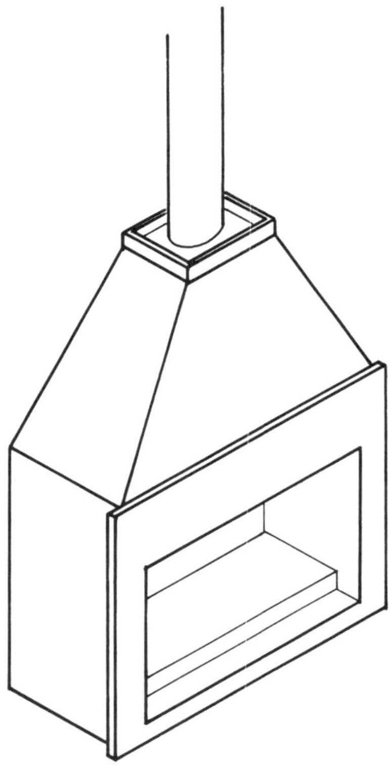

Prefabricated fireplaces are lightweight, free-standing metal units incorporating all the essential features of a built-in masonry fireplace. They generally come assembled with a heat-resistant refractory hearth to support the fire. There are also prefabricated metal chimneys available in sections to complete the installation.

Most models require some sort of heat resistant floor covering or base on which to stand, such as bricks, flagstone or gravel. Clearances for fire protection are also required although this varies with unit design and local code regulations. When installed next to a noncombustible masonry wall, this clearance may be as little as four inches, whereas 24 to 30 inches may be required for conventional wood framing.

There are many different shapes, sizes, and styles, and most come in an assortment of baked-on porcelain colors as well as black. A wide range of offset angle pieces and elbows is available to accommodate a variety of installations.

sizes

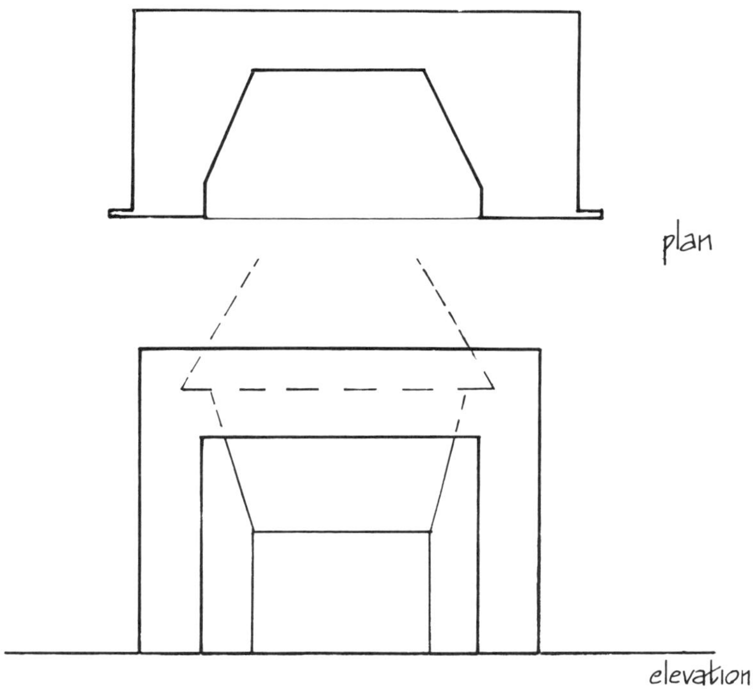

typical height: 30"

typical width: 28"
36"
42"
44"
48"

types

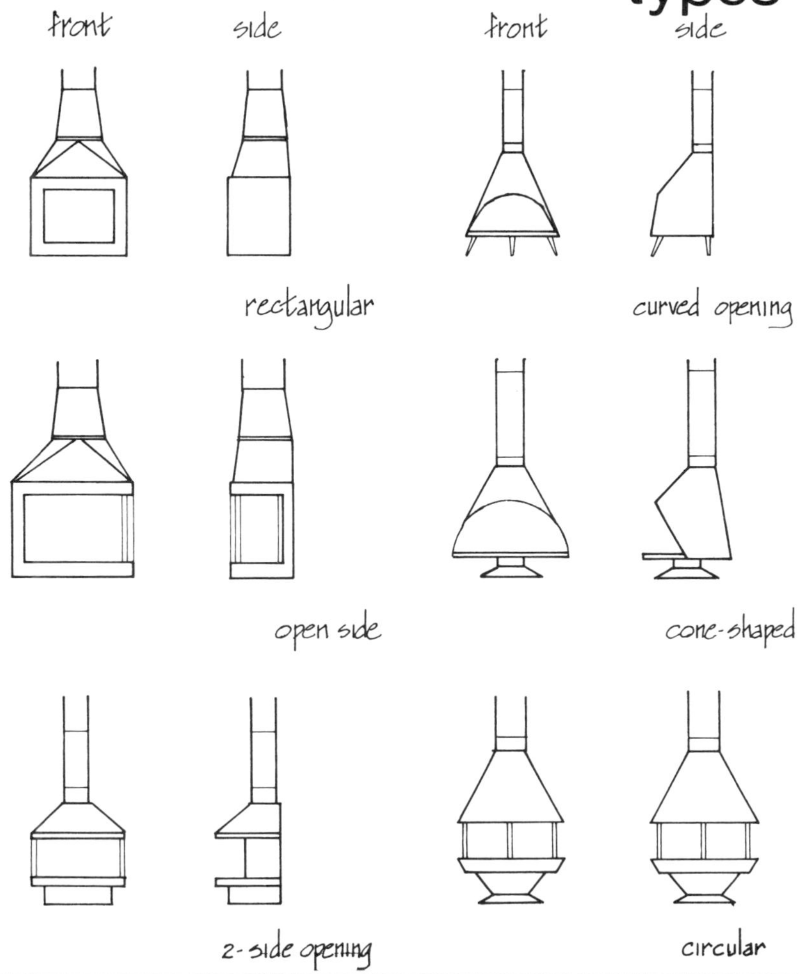

APPENDICES

ANALYSIS MATRIX

	PAGE NO.	SCALE OF PRODUCT				PRIMARY MATERIAL						DESIGN FLEX.			FABR.	
		building	system	component	element	metal	concrete	wood	other	open system	closed system	dimensional	material	finish	standard	custom
METAL BUILDING	20	●				●				●		●			●	
MODULAR CONCRETE	24	●					●				●					●
PANELIZED CONCRETE	28	●					●				●	●				●
MODULAR WOOD	32	●						●			●				●	
SECTIONAL WOOD	36	●						●			●				●	
PANELIZED WOOD	38	●						●		●		●				●
PRE-CUT WOOD	42	●						●		●		●		●		●
GLAZED BUILDING	44	●							●		●	●	●		●	
GEODESIC	48	●				●					●	●	●		●	
MOBILE HOMES	50	●				●					●				●	

MANDATORY INTERFACE																STRUC		ON-SITE LABOR				TYPE		TRANSP.			SHIPPING RADIUS			ERECT.			CONSTR. TIME		
soil	foundation	climate	structure	enclosure	hvac	plumbing	electrical	lighting/ceiling	other	codes	zoning	fire rating	acoustical rating	U-factor	relocatability	load bearing	non load bearing	manufacture	assembly	installation	finish	skilled labor	unskilled labor	conventional	special permit	by manufacturer	local	regional	national	conventional	heavy equipment	by manufacturer	excessive	conventional	efficient
	●									●	●	●		●		●			●	●	●	●		●		●		●		●				●	
	●									●	●	●	●	●		●				●		●			●	●	●				●				●
	●									●	●	●	●	●		●			●	●		●		●		●		●			●			●	
	●									●	●	●		●		●				●			●	●	●	●		●			●				●
	●									●	●	●		●		●				●			●	●	●	●			●	●					●
	●									●	●					●			●	●			●	●		●			●	●					●
	●									●	●					●			●	●	●	●		●		●			●	●				●	
	●									●	●					●			●	●		●		●		●			●	●				●	
	●									●	●	●		●		●			●	●		●		●		●			●	●				●	
										●		●			●	●				●			●	●		●			●	●					●

	PAGE NO.	SCALE OF PRODUCT				PRIMARY MATERIAL						DESIGN FLEX.			FABR.	
		building	system	component	element	metal	concrete	wood	other	open system	closed system	dimensional	material	finish	standard	custom
STEEL FRAMING	54		●			●				●		●		●	●	
STEEL STUDS	56		●			●				●		●			●	
STEEL JOISTS	60		●			●				●		●			●	
CONCRETE COLUMN & BEAM	64		●				●				●	●			●	
STRESS-SKIN PANEL	66		●					●		●		●			●	
METAL PANEL	68		●			●				●		●		●	●	
METAL CURTAIN WALL	72		●			●				●		●		●	●	
PRECAST PANEL	74		●				●				●	●		●		●
COMPOSITE PANEL	78		●						●	●		●		●	●	

MANDATORY INTERFACE																STRUC		ON-SITE LABOR				TYPE		TRANSP.			SHIPPING RADIUS			ERECT.			CONSTR. TIME		
soil	foundation	climate	structure	enclosure	hvac	plumbing	electrical	lighting/ceiling	other	codes	zoning	fire rating	acoustical rating	U-factor	relocatibility	load bearing	non load bearing	manufacture	assembly	installation	finish	skilled labor	unskilled labor	conventional	special permit	by manufacturer	local	regional	national	conventional	heavy equiptment	by manufacturer	excessive	conventional	efficient
	●		●	●						●		●				●			●	●		●		●		●		●			●			●	
			●	●						●		●				●				●			●	●		●			●	●				●	
	●		●	●						●		●				●				●			●	●		●			●	●				●	
	●			●						●		●				●				●			●	●		●		●			●			●	
			●									●	●	●		●				●			●	●		●			●	●					●
		●	●							●		●	●	●			●			●			●	●				●		●				●	
		●	●							●		●		●			●		●	●			●	●					●		●			●	
		●	●							●		●	●	●			●			●		●		●				●			●			●	
		●	●							●		●		●			●			●			●	●					●	●				●	

	PAGE NO.	SCALE OF PRODUCT				PRIMARY MATERIAL						DESIGN FLEX.			FABR.	
		building	system	component	element	metal	concrete	wood	other	open system	closed system	dimensional	material	finish	standard	custom
STEEL BAR JOISTS	82			●		●				●		●			●	
SPACE FRAME	84			●		●					●	●			●	
PRE-CAST CONCRETE	88			●			●			●		●		●	●	
PRE-CAST TEES	90			●			●			●		●			●	
CONCRETE PLANKS	92			●			●			●		●			●	
LAMINATED TIMBER	94			●				●		●		●		●	●	
WOOD TRUSSES	96			●				●		●		●			●	
METAL SIDING	100			●		●				●		●		●	●	
METAL STAIRS	104			●		●				●		●			●	
SPIRAL STAIRS	106			●		●				●		●		●	●	
CONCRETE STAIRS	108			●			●			●		●				●
SERVICE MODULE	110			●		●		●			●	●	●			●
METAL DECKING	112			●		●				●		●			●	
CONCRETE ROOFING	114			●			●			●		●			●	
WOOD ROOFING	118			●				●		●		●			●	

MANDATORY INTERFACE																STRUC		ON-SITE LABOR				TYPE		TRANSP			SHIPPING RADIUS			ERECT.			CONSTR. TIME		
soil	foundation	climate	structure	enclosure	hvac	plumbing	electrical	lighting/ceiling	other	codes	zoning	fire rating	acoustical rating	U factor	relocatibility	load bearing	non-load bearing	manufacture	assembly	installation	finish	skilled labor	unskilled labor	conventional	special permit	by manufacturer	local	regional	national	conventional	heavy equiptment	by manufacturer	excessive	conventional	efficient
			●	●	●			●				●				●				●			●	●		●		●		●				●	
			●	●	●			●				●				●			●	●		●		●					●		●				●
	●		●	●						●		●				●			●	●			●	●		●		●			●			●	
			●	●						●		●		●		●				●			●		●	●		●			●			●	
			●	●				●		●		●	●	●		●				●			●	●		●		●			●			●	
	●		●	●						●		●				●				●	●		●	●		●		●			●			●	
			●	●	●			●				●				●				●			●	●		●		●		●					●
			●											●			●			●			●	●					●	●					●
	●		●							●		●				●				●			●	●		●		●			●				●
	●									●					●	●			●	●	●		●	●					●	●					●
			●							●		●				●				●			●	●				●			●				●
				●	●	●	●			●		●	●				●			●			●		●	●		●			●				●
			●	●				●				●		●			●			●			●	●					●	●					●
		●	●	●				●		●		●		●			●			●			●	●					●	●					●
			●	●				●		●		●	●	●			●			●	●		●	●					●	●					●

	PAGE NO.	SCALE OF PRODUCT				PRIMARY MATERIAL						DESIGN FLEX.			FABR.	
		building	system	component	element	metal	concrete	wood	other	open system	closed system	dimensional	material	finish	standard	custom
DOORS	122				●	●		●		●		●	●	●	●	
WINDOWS	126				●	●		●		●		●	●	●	●	
SKYLIGHTS	130				●	●			●	●		●		●	●	
FIREPLACES	132				●	●				●		●			●	

MANDATORY INTERFACE																STRUC		ON-SITE LABOR				TYPE		TRANSP			SHIPPING RADIUS			ERECT			CONSTR. TIME		
soil	foundation	climate	structure	enclosure	hvac	plumbing	electrical	lighting/ceiling	other	codes	zoning	fire rating	acoustical rating	U-factor	relocatibility	load bearing	non-load bearing	manufacture	assembly	installation	finish	skilled labor	unskilled labor	conventional	special permit	by manufacturer	local	regional	national	conventional	heavy equiptment	by manufacturer	excessive	conventional	efficient
				●								●	●				●			●			●	●					●	●					●
				●										●			●			●			●	●					●	●					●
		●		●				●						●			●		●	●			●	●					●	●				●	
				●						●		●					●			●			●	●					●	●					●

STANDARD DATA

building live loads

	U.S. Standard (lbs./sq. ft.)	Metric Standard (kg./m²)
ASSEMBLY FACILITIES		
— theatres with fixed seats	60	293
— movable seat facilities	100	488
— corridors and lobbies	100	488
— stages	150	732
LIBRARIES		
— reading rooms	60	293
— book stacks	150	732
MANUFACTURING FACILITIES	125	610
OFFICES		
— office space	80	391
— lobbies	100	488
RESIDENTIAL FACILITIES		
— private dwellings, apartments	40	195
— public rooms	100	488
— corridors	60	293
SCHOOLS		
— classrooms	40	195
— corridors	100	488

building live loads

	U.S. Standard (lbs./sq. ft.)	Metric Standard (kg./m²)
SIDEWALKS & VEHICULAR DRIVES	250	1221
STAIRS, FIRE ESCAPES, EXITWAYS	100	488
STORAGE WAREHOUSES		
— light	125	610
— heavy	250	1221
STORES		
— retail: first floor	100	488
upper floors	75	366
— wholesale	125	610
ROOF LOADS		
— minimum, not including wind or earthquake loads	20	98
— roof-gardens	100	488

Live loads to be assumed in the design of a building should be the maximum expected to be produced by the intended use or occupancy, but never less than the above. Always verify live load requirements with local applicable building codes.

In some instances, concentrated load conditions will take precedence (e.g., office floors with heavy equipment or parking garages).

weights of materials

	U.S. Standard (lbs./sq. ft.)	Metric Standard (kg./m²)
SOIL, SAND AND GRAVEL		
— cinder	45	720.8
— clay, damp	110	1762.2
— clay, dry	63	1009.2
— earth, dry and loose	76	1217.4
— earth, moist and packed	96	1537.8
— sand and gravel, dry and loose	105	1682.1
— sand and gravel, wet	120	1922.4
WOOD		
— cedar	22	352.4
— Douglas fir	32	512.6
— hemlock	29	464.5
— maple	42	672.8
— oak, red	41	656.8
— oak, white	46	736.8
— pine, southern	29	464.5
— redwood	26	416.5
— spruce	27	432.5

	U.S. Standard (lbs./sq. ft.)	Metric Standard (kg./m²)
METAL		
— aluminum	165	2643.3
— brass	546	8746.9
— bronze, statuary	509	8154.2
— copper	556	8907.1
— iron, cast	450	7209.0
— iron, wrought	485	7769.7
— lead	710	11374.2
— monel metal	556	8907.1
— nickel	565	9051.3
— stainless steel	510	8170.2
— steel, rolled	490	7849.8
— tin	459	7351.2
— zinc	440	7048.8
CONCRETE		
— stone, plain	144	2306.8
— stone, reinforced	150	2403.0
– cinder	100	1602.0
— lightweight: haydite	85-100	1361.6-1602.0
vermiculite	25-60	400.5-961.1

	U.S. Standard (lbs./sq. ft.)	Metric Standard (kg./m²)
STONE		
— granite	175	2803.5
— limestone	165	2643.3
— marble	165	2643.3
— sandstone	147	2354.9
— slate	175	2803.5
WATER		
— max. density @ 4° C.	62	993.1
— ice	56	897.0
— snow	8	128.1
FLOOR AND ROOF CONSTRUCTION		
— concrete, reinforced, per 1″		
— stone	12.5	61
— slag	11.5	56
— lightweight	6-10	29-49
— concrete, plain, per 1″		
— stone	12	59
— slag	11	54
— lightweight	3-9	15-44
— concrete, precast hollow core plank, 6″		
— stone	40	195
— lightweight	30	146
cinder plank 2″	15	73
gypsum plank 2″	12	59
— steel deck	2-4	10-20

	U.S. Standard (lbs./sq. ft.)	Metric Standard (kg./m²)
FLOOR FINISHES		
— cement finish/topping, 1″	12	59
— marble	30	146
— terrazzo, 1″	13	63
— tile, ceramic or quarry, 3/4″	10	49
— wood:		
— hardwood 25/32″	4	20
— softwood 3/4″	2.5	12
— woodblock 3″	15	73
— vinyl asbestos tile, 1/8″	1.33	7
CEILINGS		
— acoustical fiber tile, 3/4″	1	5
— acoustic plaster on gypsum lath base	10	49
— channel suspended system	1	5
ROOFING		
— built-up (5-ply felt and gravel)	6	29
— copper or tin	6	29
— corrugated asbestos	4	20
— corrugated iron (galvanized)	1.5	7
— Fiberglas	0.5	2
— lead, 1/8″	8	39
— monel metal	1.5	7
— shingles:		
— asphalt	3	15
— asbestos cement	3	15
— slate, 1/4″	10	49
— wood	2	10

weights of materials

	U.S. Standard (lbs./sq. ft.)	Metric Standard (kg./m²)
— tile:		
— cement	16	78
— clay	14	68
WALLS AND PARTITIONS		
— brick, per 4″ of thickness	35	171
— concrete block:		
— stone and gravel agg.		
4″	34	166
6″	50	244
8″	58	283
12″	90	439
— lightweight aggregate		
4″	22	107
6″	31	151
8″	38	186
12″	55	269
— glass block, 4″	18	88
— gypsum board, 1/4″	2	10
— metal lath	0.5	2
— metal studs, with lath and plaster	18	88
— plaster, 1″		
— cement	10	49
— gypsum	5	24
— plywood, 1/2″	1.5	7
— stone:		
— granite, 4″	59	288
— limestone, 6″	55	269

	U.S. Standard (lbs./sq. ft.)	Metric Standard (kg./m²)
— marble, 1″	13	63
— sandstone, 4″	49	239
— slate, 1″	14	68
— tile:		
— ceramic mosaic, 1/4″	2.5	12
— glazed wall tile, 3/8″	3	15
— tile, structural clay		
— 4″	18	88
— 6″	28	137
— 8″	34	166
— 10″	40	195
— wood studs, 2 × 4, with lath and plaster	16	78
INSULATION		
— batt or blanket, per 1″	0.4	2
— fiber board	2	10
— foamed board, per 1″	0.2	1
— loose	0.5	2
— poured in place	2	10
— rigid	1.5	7
GLASS		
— insulating glass, with 1/8″ plate	3.25	16
— plate, 1/4″	3.30	16
— sheet:		
— (DS) 1/8″	1.60	8
— (HS) 1/4″	3.25	16
— wire glass, 1/4″	3.50	17

metric conversion factors

IMPERIAL UNIT	METRIC UNIT	CONVERSION FACTOR		
LENGTH				
mile (mi.)	kilometer (km)	1 mile	=	1.609 km
yard (yd.)	meter (m)	1 yard	=	0.9144 m
foot (ft.)	meter (m)	1 foot	=	0.3048 m
	millimeter (mm)		=	304.8 mm
inch (in.)	millimeter (mm)	1 inch	=	25.40 mm
AREA				
square mile (sq. mi.)	sq. kilmeter (km^2)	1 sq. mi.	=	2.590 km^2
	hectare (ha)		=	259.0 ha
acre (A)	hectare (ha)	1 acre	=	0.4047 ha
	square meter (m^2)		=	4046.9 m^2
square yard (sq. yd.)	square meter (m^2)	1 sq. yd.	=	0.8361 m^2
square foot (sq. ft.)	square meter (m^2)	1 sq. ft.	=	0.0920 m^2
	sq. centimeter (cm^2)		=	929.03 cm^2
square inch (sq. in.)	sq. centimeter (cm^2)	1 sq. in.	=	6.452 cm^2
VOLUME				
cubic yard (cu. yd.)	cubic meter (m^3)	1 cu. yd.	=	0.7646 m^3
cubic foot (cu. ft.)	cubic meter (m^3)	1 cu. ft.	=	0.02832 m^3
	liter (liter)		=	28.32 liter
	cubic decimeter (dm^3)		=	28.32 dm^3
cubic inch (cu. in.)	cubic millimeter (mm^3)	1 cu. in.	=	16390 mm^3
	cubic centimeter (cm^3)		=	16.39 cm^3
	millimeter (ml)		=	16.39 ml
	liter (liter)		=	0.01639 liter

metric conversion factors

IMPERIAL UNIT	METRIC UNIT	CONVERSION FACTOR
MASS		
ton (T)	kilgram (kg)	1 ton = 1016.05 kg
kip (1000 lbs.) (k)	kilogram (kg)	1 kip = 453.59 kg
pound (lb.)	kilogram (kg)	1 lb. = 0.4536 kg
ounce (oz.)	gram (g)	1 oz. = 28.35 g
CAPACITY		
quart (qt.)	liter (liter)	1 quart = 1.137 liter
pint (pt.)	liter (liter)	1 pint = 0.568 liter
fluid ounce (fl. oz.)	cubic centimeter (cm^3)	1 fl. oz. = 28.413 cm^3
MASS PER UNIT AREA		
pound/square foot (psf)	kilogram/sq. meter (kg/m^2)	1 psf = 4.882 kg/m^2
pound/square inch (psi)	kilogram/sq. meter (kg/m^2)	1 psi = 703.07 kg/m^2
ounce/square foot ($oz./ft^2$)	gram/sq. meter (g/m^2)	1 $oz./ft^2$ = 305.15 g/m^2
DENSITY		
pound/cubic foot (pcf)	kilogram/cubic meter (kg/m^3)	1 pcf = 16.02 kg/m^3
pound/cubic inch (pci)	gram/cubic centimeter (g/cm^3)	1 pci = 27.68 g/cm^3
	megagram/cubic meter (Mg/m^3)	= 27.68 Mg/m^3

metric conversion factors

IMPERIAL UNIT	METRIC UNIT	CONVERSION FACTOR
VOLUME RATE OF FLOW		
cubic feet/minute (cfm)	liter/second (liter/s)	1 cfm = 0.4791 liter/s
cubic feet/second (cfs)	cubic meter/second (m^3/s)	1 cfs = 0.02832 m^3/s
cubic inch/second (cis)	milliliter/second (ml/s)	1 cis = 16.39 ml/s
gallon/hour (gal./hr.)	liter/hour (liter/h)	1 gal/hr = 4.5461 liter/h
gallon/minute (gal/min)	liter/second (liter/s)	1 gal/min = 0.07577 liter/s
gallon/second (gal/sec)	liter/second (liter/s)	1 gal/sec = 4.5461 liter/s
FUEL CAPACITY		
gallon/mile (gpm)	liter/kilometer (liter/km)	1 gpm = 2.825 liter/km
miles/gallon (mpg)	kilometers/liter (km/liter)	1 mpg = 0.354 km/liter
VELOCITY		
miles/hour (mph)	kilometer/hour (km/h)	1 mph = 1.609 km/h
feet/minute (fpm)	meter/minute (m/min)	1 fpm = 0.3048 m/min
	meter/second (m/s)	= 0.0051 m/s
feet/second (fps)	meter/second (m/s)	1 fps = 0.3048 m/s
inch/second (ips)	millimeter/second (mm/s)	1 ips = 25.4 mm/s
TEMPTERATURE		
degree Fahrenheit (°F)	degree Celsius (°C)	1 °F = 0.5556 °C
		t °C = 0.5556 (t °F − 32)
		= $\frac{5}{9}$ (t °F − 32)

metric conversion factors

IMPERIAL UNIT	METRIC UNIT	CONVERSION FACTOR	
HEAT			
British Thermal Unit (Btu)	joule (J)	1 Btu	= 1055 J
	kilojoule (kj)		= 1.055 kJ
heat flow rate:			
B.T.U./hour (Btu/hr)	watt (W)	1 Btu/hr	= 0.2931 W
	kilowatt (kW)		= 0.0002931 kW
thermal conductivity:			
Btu•inch/ft²•hour•°F	watt/meter•deg. C (W/m²•°C)	1 Btu•inch/ft²•hr•°F	= 0.1442 W/m°C
thermal conductance:			
Btu/ft²•hour•deg F	watt/meter²•deg C (W/m²•°C)	1 Btu/ft²•hr/°F	= 5.678 W/m²•°C
refrigeration:			
ton	watt (W)	1 ton	= 3519 W
POWER			
horsepower (hp)	watt (W)	1 hp	= 745.7 W
	kilowatt (kW)		= 0.7457 kW
LIGHTING			
foot-candle (ft-candle)	lux (lux)	1 ft-candle	= 10.76 lux
lumen/square foot (lumen/ft²)	lux (lux)	1 lumen/ft²	= 10.76 lux

DECIMALS OF FOOT

1/16	0.0052	1-1/16	0.0885	2-1/16	0.1719	3-1/16	0.2552
1/8	0.0104	1-1/8	0.0938	2-1/8	0.1771	3-1/8	0.2604
3/16	0.0156	1-3/16	0.0990	2-3/16	0.1823	3-3/16	0.2656
1/4	0.0208	1-1/4	0.1042	2-1/4	0.1875	3-1/4	0.2708
5/16	0.0260	1-5/16	0.1094	2-5/16	0.1927	3-5/16	0.2760
3/8	0.0313	1-3/8	0.1146	2-3/8	0.1979	3-3/8	0.2813
7/16	0.0365	1-7/16	0.1198	2-7/16	0.2031	3-7/16	0.2865
1/2	0.0417	1-1/2	0.1250	2-1/2	0.2083	3-1/2	0.2917
9/16	0.0469	1-9/16	0.1302	2-9/16	0.2135	3-9/16	0.2969
5/8	0.0521	1-5/8	0.1354	2-5/8	0.2188	3-5/8	0.3021
11/16	0.0573	1-11/16	0.1406	2-11/16	0.2240	3-11/16	0.3073
3/4	0.0625	1-3/4	0.1458	2-3/4	0.2292	3-3/4	0.3125
13/16	0.0677	1-13/16	0.1510	2-13/16	0.2344	3-13/16	0.3177
7/8	0.0729	1-7/8	0.1563	2-7/8	0.2396	3-7/8	0.3229
15/16	0.0781	1-15/16	0.1615	2-15/16	0.2448	3-15/16	0.3281
1-0	0.0833	2-0	0.1667	3-0	0.2500	4-0	0.3333

4-1/16	0.3385	5-1/16	0.4219	6-1/16	0.5052	7-1/16	0.5885
4-1/8	0.3438	5-1/8	0.4271	6-1/8	0.5104	7-1/8	0.5938
4-3/16	0.3490	5-3/16	0.4323	6-3/16	0.5156	7-3/16	0.5990
4-1/4	0.3542	5-1/4	0.4375	6-1/4	0.5208	7-1/4	0.6042
4-5/16	0.3594	5-5/16	0.4427	6-5/16	0.5260	7-5/16	0.6094
4-3/8	0.3646	5-3/8	0.4479	6-3/8	0.5313	7-3/8	0.6146
4-7/16	0.3698	5-7/16	0.4531	6-7/16	0.5365	7-7/16	0.6198
4-1/2	0.3750	5-1/2	0.4583	6-1/2	0.5417	7-1/2	0.6250
4-9/16	0.3802	5-9/16	0.4635	6-9/16	0.5469	7-9/16	0.6302
4-5/8	0.3854	5-5/8	0.4688	6-5/8	0.5521	7-5/8	0.6354
4-11/16	0.3906	5-11/16	0.4740	6-11/16	0.5573	7-11/16	0.6406
4-3/4	0.3958	5-3/4	0.4792	6-3/4	0.5625	7-3/4	0.6458
4-13/16	0.4010	5-13/16	0.4844	6-13/16	0.5677	7-13/16	0.6510
4-7/8	0.4063	5-7/8	0.4896	6-7/8	0.5729	7-7/8	0.6563
4-15/16	0.4115	5-15/16	0.4948	6-15/16	0.5781	7-15/16	0.6615
5-0	0.4167	6-0	0.5000	7-0	0.5833	8-0	0.6667

8-1/16	0.6719	9-1/16	0.7552	10-1/16	0.8385	11-1/16	0.9219
8-1/8	0.6771	9-1/8	0.7604	10-1/8	0.8438	11-1/8	0.9271
8-3/16	0.6823	9-3/16	0.7656	10-3/16	0.8490	11-3/16	0.9323
8-1/4	0.6875	9-1/4	0.7708	10-1/4	0.8542	11-1/4	0.9375
8-5/16	0.6927	9-5/16	0.7760	10-5/16	0.8594	11-5/16	0.9427
8-3/8	0.6979	9-3/8	0.7813	10-3/8	0.8646	11-3/8	0.9479
8-7/16	0.7031	9-7/16	0.7865	10-7/16	0.8698	11-7/16	0.9531
8-1/2	0.7083	9-1/2	0.7917	10-1/2	0.8750	11-1/2	0.9583
8-9/16	0.7135	9-9/16	0.7969	10-9/16	0.8802	11-9/16	0.9635
8-5/8	0.7188	9-5/8	0.8021	10-5/8	0.8854	11-5/8	0.9688
8-11/16	0.7240	9-11/16	0.8073	10-11/16	0.8906	11-11/16	0.9740
8-3/4	0.7292	9-3/4	0.8125	10-3/4	0.8958	11-3/4	0.9792
8-13/16	0.7344	9-13/16	0.8177	10-13/16	0.9010	11-13/16	0.9844
8-7/8	0.7396	9-7/8	0.8229	10-7/8	0.9063	11-7/8	0.9896
8-15/16	0.7448	9-15/16	0.8281	10-15/16	0.9115	11-15/16	0.9948
9-0	0.7500	10-0	0.8333	11-0	0.9167	12-0	1.0000

DECIMALS OF INCH

1/64	0.015625	17/64	0.265625	33/64	0.515625	49/64	0.765625	9/64	0.140625	25/64	0.390625	41/64	0.640625	57/64	0.890625
1/32	0.03125	9/32	0.28125	17/32	0.53125	25/32	0.78125	5/32	0.15625	13/32	0.40625	21/32	0.65625	29/32	0.90625
3/64	0.046875	19/64	0.296875	35/64	0.546875	51/64	0.796875	11/64	0.171875	27/64	0.421875	43/64	0.671875	59/64	0.921875
1/16	0.0625	5/16	0.3125	9/16	0.5625	13/16	0.8125	3/16	0.1875	7/16	0.4375	11/16	0.6875	15/16	0.9375
5/64	0.078125	21/64	0.328125	37/64	0.578125	53/64	0.828125	13/64	0.203125	29/64	0.453125	45/64	0.703125	61/64	0.953125
3/32	0.09375	11/32	0.34375	19/32	0.59375	27/32	0.84375	7/32	0.21875	15/32	0.46875	23/32	0.71875	31/32	0.96875
7/64	0.109375	23/64	0.359375	39/64	0.609375	55/64	0.859375	15/64	0.234375	31/64	0.484375	47/64	0.734375	63/64	0.984375
1/8	0.125	3/8	0.375	5/8	0.625	7/8	0.875	1/4	0.250	1/2	0.500	3/4	0.750	1-0	1.000

TRADE ASSOC.

AFL-CIO	Building and Construction Trades Department 815 16th St., N.W. Washington, D.C. 20006
Aluminum Association	750 Third Avenue New York, N.Y. 10017
American Building Contractors Association	3345 Wilshire Blvd. Los Angeles, California 90010
American Concrete Institute	P.O. Box 4754 Redford Station Detroit, Michigan 48219
American Forest Institute	1619 Massachusetts Ave., N.W. Washington, D.C. 20036
American Institute of Architects	1785 New York Ave., N.W. Washington, D.C. 20036
American Institute of Building Design	Union Bank Plaza, Suite 408 15233 Ventura Blvd. Sherman Oaks, California 91403
American Institute of Constructors	4545 N. Lincoln Blvd. Suite 298 Oklahoma City, Oklahoma 73105
American Institute of Planners	917 15th St., N.W. Room 800 Washington, D.C. 20005

American Institute of Steel Construction, Inc.	101 Park Ave. New York, N.Y. 10017
American Institute of Timber Construction	333 W. Hampden Ave. Englewood, Colorado 80110
American Iron and Steel Institute	150 E. 42nd St. New York, N.Y. 10017
American National Standards Institute	1430 Broadway New York, N.Y. 10018
American Plywood Institute	1119 A St. Tacoma, Washington 98401
American Savings and Loan Institute	111 E. Wacker Drive Chicago, Illinois 60601
American Society for Testing and Materials	1916 Race St. Philadelphia, Pennsylvania 19103
American Society of Concrete Constructors	2510 Dempster St. Des Plains, Illinois 60016
American Society of Civil Engineers	345 E. 47th St. New York, N.Y. 10017
American Society of Planning Officials	1313 E. 60th St. Chicago, Illinois 60637
Architectural Aluminum Manufacturers Association	410 N. Michigan Ave. Chicago, Illinois 60611
Associated General Contractors of America, Inc.	1957 E. St., N.W. Washington, D.C. 20006
Battelle Memorial Institute	505 King Ave. Columbus, Ohio 43201
Bemis Foundation	Massachusetts Institute of Technology Cambridge, Massachusetts 02142
Building Officials and Code Administrators International	1313 E. 60th St. Chicago, Illinois 60637
Building Research Institute	2101 Constitution Ave., N.W. Washington, D.C. 20418
Building Systems Information Clearinghouse	Educational Facilities Laboratory 3000 Sand Hill Road Menlo Park, California 94025
California Redwood Association	617 Montgomery St. San Francisco, California 94111
Council of Housing Producers	1801 Avenue of the Stars Los Angeles, California 90067
Construction Specifications Institute, Inc.	1717 Massachusetts Ave., N.W. Washington, D.C. 20036
Division of Building Research of Canada	Montreal Road Ottawa, 7, Canada

Facing Tile Institute	111 E. Wacker Drive Chicago, Illinois 60601
Federal Housing Administration	451 7th St., S.W. Washington, D.C. 20411
Forest Products Laboratory	U.S. Department of Agriculture Madison, Wisconsin 53705
Gypsum Association	201 N. Wells St. Chicago, Illinois 60606
IIT Research Institute	10 W. 35th St. Chicago, Illinois 60616
International Conference of Building Officials	50 S. Los Robles Pasadena, California 91101
Metal Building Manufacturers Association	2130 Keith Building Cleveland, Ohio 44115
Mobile Homes Manufacturers Association	P.O. Box 201 14650 Lee Road Chantilly, Virginia 22021
National Association of Architectural Metal Manufacturers	1010 W. Lake St. Oak Park, Illinois 60301
National Association of Building Manufacturers	1619 Massachusetts Ave., N.W. Washington, D.C. 20036
National Association of Home Builders	1625 L. St., N.W. Washington, D.C. 20036
National Association of Housing and Redevelopment Officials	2600 Virginia Ave., N.W. Washington, D.C. 20037
National Bureau of Standards	U.S. Department of Commerce Washington, D.C. 20234
National Concrete Masonry Association	1800 N. Kent St. Arlington, Virginia 22209
National Forest Products Association	1619 Massachusetts Ave., N.W. Washington, D.C. 20036
National Hardwood Lumber Association	59 E. Van Buren St. Chicago, Illinois 60605
National Home Improvement Council	11 E. 44th St. New York, N.Y. 10017
National Housing Center	15th and M Sts., N.W. Washington, D.C. 20005
National Housing Conference	1250 Connecticut Ave., N.W. Washington, D.C. 20036
National Particleboard Association	2306 Perkins Place Silver Springs, Maryland 20910
National Woodwork Manufacturers Association, Inc.	400 W. Madison St. Chicago, Illinois 60606
Perlite Institute, Inc.	45 W. 45th St. New York, N.Y. 10036

Pierce Foundation	Housing Research Division New York, N.Y.
Portland Cement Association	5420 Old Orchard Road Skokie, Illinois 60076
Prestressed Concrete Institute	20 N. Wacker Drive Chicago, Illinois 60606
Producers Council	1717 Massachusetts Ave., N.W. Washington, D.C. 20036
Society of American Registered Architects	111 E. Wacker Drive Chicago, Illinois 60601
Southern Building Code Congress	1116 Brown-Marx Building Birmingham, Alabama 35203
Southern Forest Products Association	3525 Causeway Blvd. New Orleans, Louisiana 70150
Stanford Research Institute	Stanford University Menlo Park, California 94025
Steel Door Institute	2130 Keith Building Cleveland, Ohio 44115
Steel Joist Institute	2001 Jefferson Davis Highway Arlington, Virginia 22202
Steel Window Institute	2130 Keith Building Cleveland, Ohio 44115
Structural Clay Products Institute	1750 Old Meadow Road McLean, Virginia 22101
Superintendent of Documents	U.S. Government Printing Office Washington, D.C. 20402
Trailer Coach Association	3855 E. La Palma Ave. Anaheim, California 92806
Underwriters' Laboratories, Inc.	207 E. Ohio St. Chicago, Illinois 60611
U.S. Department of Housing and Urban Development	451 7th St., S.W. Washington, D.C. 20410
United States Savings and Loan League	111 E. Wacker Drive Chicago, Illinois 60601
Urban Land Institute	1200 18th St., N.W. Washington, D.C. 20036
Vermiculite Institute	141 W. Jackson Blvd. Chicago, Illinois 60604
Western Wood Products Association	1500 Yeon Building Portland, Oregon 97204

GLOSSARY

Assembly	A combination of building parts into a clearly identifiable, specific aspect of the building.
Basic Module	A coordinating, dimensioning unit for systems, components, and elements.
Building Component	See Component.
Building Process	The complete process of building from conception to total satisfaction of requirements.
Building Subsystem	See Subsystem.
Building System	A set of interrelated and coordinated building parts which function as a total entity to satisfy some or all of the functions of a building.
Closed System	A building system which only permits interchangeability of its own subsystems and components.
Compatibility	The state of dimensional, functional, or other coordination between two or more building subsystems or components; see Interface.

Component	A prefabricated part of a building which is manufactured and considered as an independent unit capable of being combined with other building components and elements to produce the whole building.
Design Process	A sequence of decisions involving the transformation of the building requirements (program) into physical form capable of being built.
Dimensional Coordination	A system of organizing the dimensions of systems, components, and elements so that they may be combined and assembled without any need for modification.
Element	The smallest or least complex fabricated item which has an identity of its own.
Functional Coordination	A system of organizing the performance characteristics of prefabricated parts to allow optimum performance in their use and combination.
Functional Element	A part of a building which can be distinguished from other parts by the function it satisfies.
Hardware	The physical parts of a system, component, or element.
H.U.D.	US Department of Housing and Urban Development.
Industrialization	The aspect of the manufacturing process through which, by the use of technological, mass production techniques, there is an increase in productivity, efficiency, or quality of the product.
Industrialized Building	Building incorporating industrialized parts.
Industrialized Building System	The modification of the building process to allow the total integration of industrialized production, transportation, and assembly techniques.
Interface	The juncture or boundary between two building systems, components, or elements.

Mobile Home	A portable structure built on a frame or chassis, and designed to be easily transported or relocated.
Modular	Pertaining to something having dimensions based upon a module or multiples thereof.
Modular Coordination	A system of sizing the dimensions of parts to allow easy and standard combinations and applications.
Module	A basic fixed dimension; also, a three-dimensional unit with specific dimensions.
Open System	A building system which permits interchangeability of its own subsystems, components, and elements with others.
Parts	See Element
Performance Criteria	A statement of the essential characteristics that a building or part of a building must perform to satisfy user needs.
Performance Specifications	A description of a building system, component, or element based upon its desired performance, not upon its physical characteristics.
Planning Grid	A reference grid, usually rectangular, with the spacing of the grid lines determined by the module of the building system, which serves as a dimensional framework for organizing the building.
Prefabrication	The assembly of certain parts of the final building prior to its delivery to the job site.
Rationalized Building	Conventional building with the adaptation of some of the principles and techniques of systems building.
Software	The nonphysical parts of a system, such as a program or procedure.
Subsystem	A portion of a total system which performs in a specific manner.

System — A kit of parts designed to be combined into a unified whole, to accomplish a desired objective.

Systems Approach — The viewing of a problem as a set of interrelated, interdependent parts which work together for the overall objectives of the whole; also referred to as Systems Analysis.

Systems Building — The application of systems approach to construction, organizing the various facets of the building process into an efficient total process.

Tolerance — An allowance for dimensional variations within acceptable limits to retain proper performance or connection to other parts.

U-Factor — A coefficient of heat loss.

Unitized Construction — A type of systems building in which three-dimensional units (usually boxes) form assemblies.

User Needs — Those conditions which the user of a building considers necessary or desirable as environment and support for his activities, without particular reference to how such conditions are to be physically produced.

BIBLIOGRAPHY

"The aesthetics and technology of preassembly." *Progressive Architecture,* Oct., 1964, p. 162–222.

A series of articles discussing industrialized building present and proposed, in Europe and in the U.S. Descriptions of 16 companies' systems.

Alonso, William

Innovation in Housing Design and Construction Techniques as Applied to Low-Cost Housing. A collateral literature survey. Prepared for Kaiser Engineers by William Alonso and others. Berkeley, California, 1969.

A bibliography of more than a thousand items, prepared in connection with Phase 2 of the In-Cities Experimental Housing Research and Development Project of the Department of Housing and Urban Development.

Bemis, Albert Farwell

The Evolving House, (Rational Design, vol. 3), Technology Press, Cambridge, Massachusetts, 1933.

A classic work on housing and industrialization.

Bender, Richard

A Crack in the Rear-View Mirror: A View of Industrialized Building, Van Nostrand Reinhold Company, New York, 1973.

"This book surveys the current state-of-the-art of the building industry.... The author views [the general public's] obsolete image of both housing and industrialization as a major roadblock to the growth of industrialized building. He demonstrates how building construction can be better adapted to the social condition of our day as well as to contemporary means of production."

Bender, Richard
Selected Technological Aspects of the American Building Industry. The Industrialization of Building.
Prepared for the National Commission on Urban Problems. New York, 1968.

Bishop, D.
System Building In Europe. Garston, England, Building Research Station, Ministry of Technology, 1966.
Critical analysis of the factors operating behind the scenes of the building industries in Sweden, France and the U.S.S.R.

Brandle, Kurt
The Systems Approach to Building, A Core of Information. ACSA Learning Package, Graduate School of Architecture, University of Utah, Parts 1 and 2, 1974; Part 3, 1976.
A clear presentation of the subject, designed to be a resource for schools of architecture.

Bruce, Alfred and Sendbank, Harold
A History of Prefabrication, Arno Press, New York, 1972.
"This publication, which discusses systems and developments in the prefabrication movement, is based on a more extensive research study for the Pierce Foundation."

Building Research Advisory Board, Special Advisory Committee on Industrialized Housing and Building Systems, National Academy of Sciences-National Research Council.
"An historical evaluation of industrialized housing and building systems in the United States." *In* U.S. President's Committee on Urban Housing Report, Technical Studies, Vol. I, p. 177–189.
"While industrialization in housing has been progressing over the years, and while the technical knowledge now exists, application has been limited by the 'building envelope' which represents only 1/6 of the cost."

Building Systems Information Clearinghouse
Publications:
BSIC Special Report No. 1: Manufacturer's Compatibility Study (1971)
BSIC Special Report No. 3: Building Systems Planning Manual (1971)
BSIC Research Report No. 2: Evaluation, Two Studies of SCSD Schools (1972)
BSIC Research Report No. 3: A History and Evaluation of the SCSD Project, 1961–67 (1971)

Carreiro, Joseph
The New Building Block, A Report on the Factory-Produced Dwelling Module, Research Report No. 8, Cornell University, Ithaca, New York, 1968.
"A Review of what is happening in the field of modular construction, with a brief commentary on individual efforts and their possible significance for the future of housing."

Central Mortgage and Housing Corp.
Catalog of House Buliding Construction Systems, Ottawa, 1960.
Compendium of known construction methods of single family dwellings.

Cherner, Norman
Fabricating Houses from Component Parts, Reinhold Publishing Corp., N.Y., 1957.
"Economic building requires the use of readily available components. These are becoming more numerous, are more simple to assemble, and can be installed more easily.... [the author] has compiled in this book a series of basic designs, giving details, and materials listings..."

Cutler, Laurence and Cutler, Sherrie
Handbook of Housing Systems for Designers and Developers, Van Nostrand Reinhold, N.Y., 1974.
"Describes how to evaluate specific housing systems and explains building systems as a process. Graphics and plans provide a 'kit of parts' to aid the reader in planning modular housing system projects."

Diamant, R.M.E.
Industrialized Buildings, Volumes I, II, III, Iliffe Books Ltd., London, 1968.
Published under the auspices of the Architect and Building News. A survey of European systems.

Dietz, Albert G.H. and Cutler, Laurence, editors
Industrialized Building Systems for Housing, M.I.T. Press, Cambridge, Massachusetts, 1971.
"Based on two special summer programs held at M.I.T., [this book] contains contributions from experts in areas ranging from urban politics to systems analysis and materials technology. It offers a strict examination of the underlying principles involved in industrialization and the types of building systems now evolving, particularly in housing."

Educational Facilities Laboratory
SCSD: The Project and the Schools; A Report from EFL, N.Y., 1967.
History of the SCSD project and the finished schools.

Ehrenkrantz, E.D.
"The system to systems", *AIA Journal,* May 1970, p. 56–59.
Discussion of the concept and terminology of system building.

"Factory-Built Houses — solution for the shelter shortage?", *Morgan Guaranty Survey,* Aug., 1969, p. 6–11.
Discusses the problems causing the housing shortage and present methods of housing construction. Suggests that industrialized housing will help alleviate the crisis.

Grubb, C.A. and Phares, M.I.
Industrialization: A New Concept for Housing, Praeger Publishers, N.Y., 1972.
"Chapters in the book cover housing today, the changing housing consumer, the financing of housing, land and site development, housing construction, the changing product, technology in housing, housing and the community, rehabilitation and maintenance, the new town of Jonathan, Minnesota, and industrialization of housing in terms of the concept and as a business venture."

Guy, R.B.
The State of the Art of Prefabrication in the Construction Industry. Final report to the Building and Construction Trades Dept., AFL-CIO. Columbus, Ohio, Battelle Memorial Institute, 1967.
"Summarizes prefabrication experience in Europe and the U.S. Special attention is given to the probable constraints on future growth of prefabrication and its impact on the various building trades."

Habraken, N.J.
Supports: An Alternative to Mass Housing, Praeger Publishers, N.Y., 1972.
By considering the different needs of the various systems involved in housing, a more flexible solution to the problem may be arrived at.

Hardless, Trevor, editor
Europrefab Systems Handbook: Housing, London, Interbuild Prefabrication Publications, Ltd., 1969.
Europrefab, an international federation of national associations for the promotion of prefabricated and other industrialized building systems, has published an extensive and detailed survey of the building systems used in twelve countries.

Hodes, Daniel A. and Jensen, Gordon F.
The Case for Industrialized Housing — Reexamined. Housing Research Incorporated, Fairfield, Connecticut, 1973.
"The need to reexamine industrialized housing and to reassess its true potential is urgent, primarily for two reasons: (1) to measure the impact of the recent shakeout of key producers of industrialized housing, particularly in the modular housing segment, and (2) to clear up misunderstandings regarding the entire area of industrialized housing... It is hoped that the study will assist present and future housing industry participants to gain perspective on industrialized housing, particularly its market potential and requirements for successful participation or entry."

Hooper, William L.
New Housing Systems Concepts. Chicago, Mobile Home Manufacturers Association, 1967.
Leaders of the mobile home manufacturing industry present their views on the potential of the mobile home concept, i.e., factory constructed sectionalized living units as a solution to the problems of the shortage of satisfactory housing for low- and middle-income residents of urban areas.

"Housing as process", *Architectural and Engineering News,* Aug., 1968, p. 18-19.
"Progress in housing today is hobbled by viewing a house as a custom-made product using highly paid, skilled craftsmen, with costly-to-maintain-and-replace components. What is more, house construction is financed under a system which makes it the first casualty in any shift in the American economy."

Illinois Institute of Technology
Research Institute Application of Component Construction to Multi-story, Low-income Housing; final report, Chicago, 1968.
Description of planning and production of three dimensional concrete room-size modules by factory methods.

"Industrialized building", *Professional Builder,* Aug., 1969, p. 41-48.
A special section devoted to manufactured housing systems and components.

Interbuild
System Building. London, annual.
Publishes the latest developments by architects, engineers, builders and manufacturers toward the rationalization of technique and design from housing, schools, universities and hospitals down to individual components.

International Council for Building Research, Studies and Documentation
Innovation in Building, Contributions at the Second CIB Congress, Cambridge, 1962. Amsterdam, Elsevier, 1962.
Presents excellent background material for understanding industrialized building.

International Council for Building Research, Studies and Documentation
Towards Industrialized Building; proceedings of the Third CIB Congress, Copenhagen, 1965. Amsterdam, New York, Elsevier Publishing Co., 1966.
Gives a clear picture of the sophistication of industrialized building research throughout Europe.

International Systems Building Roundtable Conference, Synopsis and Proceedings, Boston Architectural Center, 1973.
A series of lectures and discussions held at the Boston Architectural Center under the topic of system building.

Jones and Laughlin Steel Corp.
New Concepts in Low-Cost Residential Housing. Pittsburgh, 1967.
Presents seven different sectionalized housing designs.

Kelly, Burnham
The Prefabrication of Houses; a study by the Albert Farwell Bemis Foundation of the prefabrication industry in the United States. New York, Wiley, 1951.
A classic work on prefabrication. Gives a complete history of the prefabrication industry and deals with such topics as management, design, procurement, production and marketing.

LCN Newsletter
Systems building in Canada, parts 1-6, 1971-1973.
A series of articles on systems building.

"Lowering the cost of housing: the meandering path to six million homes." *Progressive Architecture,* June, 1968, p. 94-139.
A thorough series of articles discussing the need for mass housing, the obstacles, the role of the Federal government, especially HUD, technological advances in prefabrication, mobile homes, turnkey housing, and sociological implications of low-cost housing.

Luder, Owen
"In search of a kit of parts." *Building,* Nov., 1968, p. 75-76.
Discusses the basic structure of the building industry; labor costs; progress in industrialization; system building; the

reaction against closed systems; development of the kit-of-parts approach; and cooperation between architect and industrial designer.

National Association of Home Builders
A Compendium of Building Systems and Components, edited by Benjamin Shapiro, Washington, 1969.

National Housing Center
Industrialized Building and Related Topics; a selected list of periodical articles published in 1968 and 1969, and including books published 1965-1969, Washington, 1970.

"Open system U.S.A." *Architectural and Engineering News,* June, 1967, p. 70-73.
A catalog of components for structure, enclosure, roofs, floors, ceilings, partitions, communications, controls, bathrooms, etc.

"Pre-engineered buildings." *Architectural and Engineering News,* Sept., 1966, p. 59-61.
Describes a number of prefabricated systems available. Tables list typical metal buildings and typical building systems with information on materials, spans, cost, delivery time, type of labor required.

"Pre-fab: what's in it for U.S. Development programs?", *Journal of Housing,* Sept., 1966, p. 435-447.
Details pre-fab experiences in various locations to indicate that in certain places prefabricated systems may solve certain problems, but implies that the current development of techniques would be insufficient to solve the housing dilemma.

Testa, Carlo
The Industrialization of Building, New York, Van Nostrand Reinhold, 1972.
"This work contains a description of the industrialization of building as a total process, with due consideration being taken of the functions of the various participants in the whole system."

Toronto School Board
Study of Education Facilities:
SEF T1: Introduction to the first SEF building system
SEF T2: Specifications for the first SEF building system. Toronto, 1968.
Details the Canadian system school building.

United Nations
Industrialization of Building. New York, 1967.
Covers the problem of industrialization in developing countries.

U.S. Department of Housing and Urban Development
Feedback: Operation Breakthrough, vol. 1, 2, 3. Government Printing Office, Washington.
Volume 1: Design and Development of Housing Systems for Operation Breakthrough, 1973.
"This document is a compendium of the 22 housing sys-

tems designed for Operation Breakthrough"

Volume 2: A Compendium of Building Concepts, 1974.

"This Compendium describes some of the innovative concepts and techniques which have been developed and utilized by the Housing System Producers selected by HUD to participate in the program."

Volume 3: A Documentary of Transportation and Handling Systems, 1974.

"The purpose of this document is to provide a general record of the transportation, handling, storage and erection systems employed by each of the Breakthrough producers ... and to describe certain methods that appear to offer opportunities in reducing overall logistics and construction costs while fully protecting the finished module, panel or component from damage."

U.S. Department of Housing and Urban Development

Housing Systems Proposals for Operation Breakthrough, Government Printing Office, Washington, 1971.

"The resource of Operation Breakthrough housing systems proposals represents an aggregate view of the technical state-of-the-art, and therefore this compilation of proposals is of value to all those contributing to the advancement of residential development."

U.S. Department of Housing and Urban Development

Operation Breakthrough, Mass Produced and Industrialized Housing, A Bibliography. Government Printing Office, Washington, 1970.

An extensive bibliography of books and articles on the subject of industrialization.

Wachsmann, Konrad

The Turning Point in Building, New York, Reinhold Publishing Corp., 1961.

A classic work by one of the early leaders in industrialization and systems.

PERIODICALS

American Institute of Architects Journal	1735 New York Ave., N.W. Washington, D.C. 20006
Apartment Construction News	Gralla Publications 7 E. 43rd St. New York, N.Y. 10017
Architectural Design	Standard Catalogue Co., Ltd. 26 Bloomsbury Way London WC1, England
Architectural Record	McGraw-Hill, Inc. P.O. Box 430 Hightstown, N.J. 08520
Automation in Housing	Vance Publishing Corporation 300 West Adams St. Chicago, Illinois 60606
Build International	Weena 700 Rotterdam, Netherlands
Building	The Builder House 4, Catherine Street London, WC2, England
Building Design and Construction	Industrial Publications, Inc. 5 S. Wabash Ave. Chicago, Illinois 60603
Building Research: dedicated to the advancement of the science and technology of building	BRAB, Building Research Institute National Academy of Sciences 2101 Constitution Ave., N.W. Washington, D.C. 20418
Building Research News	Division of Building Research National Research Council Ottawa, 7, Canada
Building Standards	International Conference of Building Officials 50 S. Los Robles Ave. Pasadena, California 91101
Building Systems Design	Presstech Design, Inc. 235 Duffield St. Brooklyn, N.Y. 11201
Construction Specifier	Construction Specifications Institute 1717 Massachusetts Ave., N.W. Washington, D.C. 20036
Constructor	Associated General Contractors of America 1957 E St., N.W. Washington, D.C. 20006

Periodical	Address
Engineering News-Record	McGraw-Hill Publications 330 W. 42nd St. New York, N.Y. 10036
House & Home	McGraw-Hill, Inc. 330 W. 42nd St. New York, N.Y. 10036
Industrial Design	Whitney Publications, Inc. 18 E. 50th St. New York, N.Y. 10022
Industrialization Forum	School of Architecture Box 1079 Washington University St. Louis, Missouri 63130
Journal of Housing	National Association of Housing and Redevelopment Officials 1413 K St., N.W. Washington, D.C. 20005
NAHB Journal of Home Building	National Association of Homebuilders 1625 L St., N.W. Washington, D.C. 20036
Professional Builder	Cahners Publishing Co. 5 S. Wabash Ave. Chicago, Illinois 60603
Progressive Architecture	Reinhold Publishing Corporation 614 Superior Ave. W. Cleveland, Ohio 44113
Sweet's Catalog Files	McGraw-Hill, Inc. 330 W. 42nd St. New York, N.Y. 10036
Systems Building News	1760 Peachtree Road, N.W. Atlanta, Georgia 30309
Technical Bulletin and Review of Quality Building Products	Producers Council 1717 Massachusetts Ave., N.W. Washington, D.C. 20036
Technology Review	Alumni Association of The Massachusetts Institute of Technology Cambridge, Massachusetts 02139

INDEX